A Collection of Contemporary Toilet Designs

EOOS and WEDC

In 2011, the Water, Sanitation & Hygiene program at the Bill & Melinda Gates Foundation initiated the Reinvent the Toilet Challenge to bring sustainable sanitation solutions to the 2.5 billion people worldwide who don't have access to safe, affordable sanitation.

Grants have since been awarded to researchers and industries around the world who are using innovative approaches – based on fundamental engineering processes – for the safe and sustainable management of human waste. The Reinvent the Toilet Challenge aims to create a toilet that:

- Removes germs from human waste and recovers valuable resources such as energy, clean water, and nutrients.
- Operates 'off the grid' without connections to water, sewer, or electrical lines.
- Costs less than US$.05 cents per user per day.
- Promotes sustainable and financially profitable sanitation services and businesses that operate in poor, urban settings.
- Is a truly aspirational next-generation product that everyone will want to use – in developed as well as developing nations.

Innovative solutions change people's lives for the better. By applying creative thinking to everyday challenges, such as dealing with human waste, we can fix some of the world's toughest problems. We hope the designs displayed in this book will encourage further research and investments in improved sanitation.

Doulaye Kone, PhD
Senior Program Officer, WSH, Toilet Team
Transformatives Technologies
Global Development

Water, Engineering and Development Centre,
Loughborough University,
Leicestershire, LE11 3TU, UK

© WEDC/Practical Action Publishing Ltd, 2014

First published in 2014

ISBN: 978 1 84380 155 9

All rights reserved. No part of this publication may be reprinted or reproduced
or utilized in any form or by any electronic, mechanical, or other means, now
known or hereafter invented, including photocopying and recording,
or in any information storage or retrieval system,
without the written permission of the publishers.

A catalogue record for this book is available from the British Library.

WEDC (The Water, Engineering and Development Centre) at Loughborough
University in the UK is one of the world's leading institutions concerned with
education, training, research and consultancy for the planning, provision and
management of physical infrastructure for development in low- and middle-
income countries.

This edition reprinted and distributed by Practical Action Publishing.

Since 1974, Practical Action Publishing has published and disseminated books
and information in support of international development work throughout the
world. Practical Action Publishing trades only in support of its parent charity
objectives and any profits are covenanted back to Practical Action
(Charity Reg. No. 247257, Group VAT Registration No. 880 9924 76).

Book edited and designed by Rod Shaw
Illustrations by Ken Chatterton

An online copy of this publication is available from:
http://wedc.lboro.ac.uk/knowledge/

Hypertext links to further information listed in this volume were active
as of 18 January 2014.

All reasonable precautions have been taken by the WEDC, Loughborough University to verify
the information contained in this publication. However, WEDC, Loughborough University does
not necessarily endorse the technologies presented in this document. The published material is
being distributed without warranty of any kind, either expressed or implied. The responsibility
for the interpretation and use of the material lies with the reader. In no event shall the WEDC,
Loughborough University be liable for damages as a result of their use.

Preface

In August 2012, I met the EOOS representative in Seattle through Dr Carl Hensman of the Bill & Melinda Gates Foundation where we started to think about ways to present a collection of contemporary toilet designs that would spark a wider interest in the subject.

Alongside the Reinvent the Toilet Challenge of the Foundation, EOOS (based in Austria) had started to collect information relating to the design of its Blue Diversion Toilet as well as other toilet designs. The information was initially gathered together in a research log, which, although not a project with an encyclopaedic scientific focus, provided insight into new and innovative ways of addressing global sanitation issues.

This collection is the result of the findings of EOOS research which was supported by Sandec, the Department of Water and Sanitation in Developing Countries at the Swiss Federal Institute of Aquatic Science and Technology (Eawag). It covers a wide range of contemporary designs along with a valuable list of website links where additional information about each design can be sought.

The research log was an interesting tool for the transfer of knowledge internally. Externally, designers, manufacturers, social researchers and others, including the Gates Foundation, appreciated the work as it served as a point of departure for toilet design projects for those new to the field.

This volume is a synthesis of the initial research log, designed and produced by The Water Engineering and Development Centre (WEDC) at Loughborough University. As conventional toilet designs are not included, it does not claim to be fully comprehensive but it nevertheless provides a useful overview of current research and development for fieldworkers and practitioners as well as engineers and researchers.

As a member of WEDC, Loughborough University, and a lead on one of the RTTC grants, I am pleased that this document is now available for a wider audience. But this is only a start: we intend to keep adding new innovations to our collection. So if there are any innovations we have missed please let us know. I sincerely hope that this endeavour will help to make the toilet a more desirable product, one that people will cherish in their homes.

Professor M.Sohail Khan
BEng, MSc, PhD Loughborough, Fellow ASCE (USA)
Professor of Sustainable Infrastructure
Water, Engineering and Development Centre (WEDC)
School of Civil and Building Engineering
Loughborough University

A number of the designs are the result of student projects and these are marked with an asterisk (*).

Contents

Contemporary Toilet Designs 1

2P Portable Restroom* 2

Aquatron Toilet 3

Blue Diversion Toilet 4

Bipee* 5

Built-in Composting System 6

Camping Toilets 7

Caravan Toilets 8

Cinderella Toilet 9

Drysan Waterless Toilet 10

EkoToi 11

Envirosan 12

Eram Delight 13

Fresh Life Toilet 14

Ghanasan 15

Green, Portable Eco-toilet* 16

Hightech Composting Toilet: Ecodomeo 17

IDE EZ Latrine 18

Indoor Self-contained Toilet 19

Intestinal Toilet 20

Locus Toilet 21

Loowatt Toilet 22

MoSan – Mobile Sanitation 23

Open Privy* 24

Otji Toilet 25

PeePoo 26

Piet Urine Diversion Toilet* 27

Public Rest Room 28

Resource Toilet 29

Rolling Toilet: x-runner 30

SimSan 31

Waterless Toilet* 32

WC der Zukunft 33

WooWoo 34

Controlling Odour 35

Controlling odour from urine 36

    Rubber tube seal 36

    Condoms – The low-tech alternative 36

    Valve seals..36

    Sealant liquid..37

    Light bulb / table tennis ball 37

Controlling odour from faeces 38

    Ventilation..38

    Soil, ash, sawdust, sand or lime...38

Contemporary Toilet Designs

Contemporary Toilet Designs

2P Portable Restroom*

The 2P is a portable restroom designed to counter the problems that arise due to heavy attendance at outdoor events. Through innovative placement of an external urinal, it promotes dual-functionality per unit which effectively cuts all lines in half, improves hygienic conditions and reduces cost, space, and it's carbon footprint.

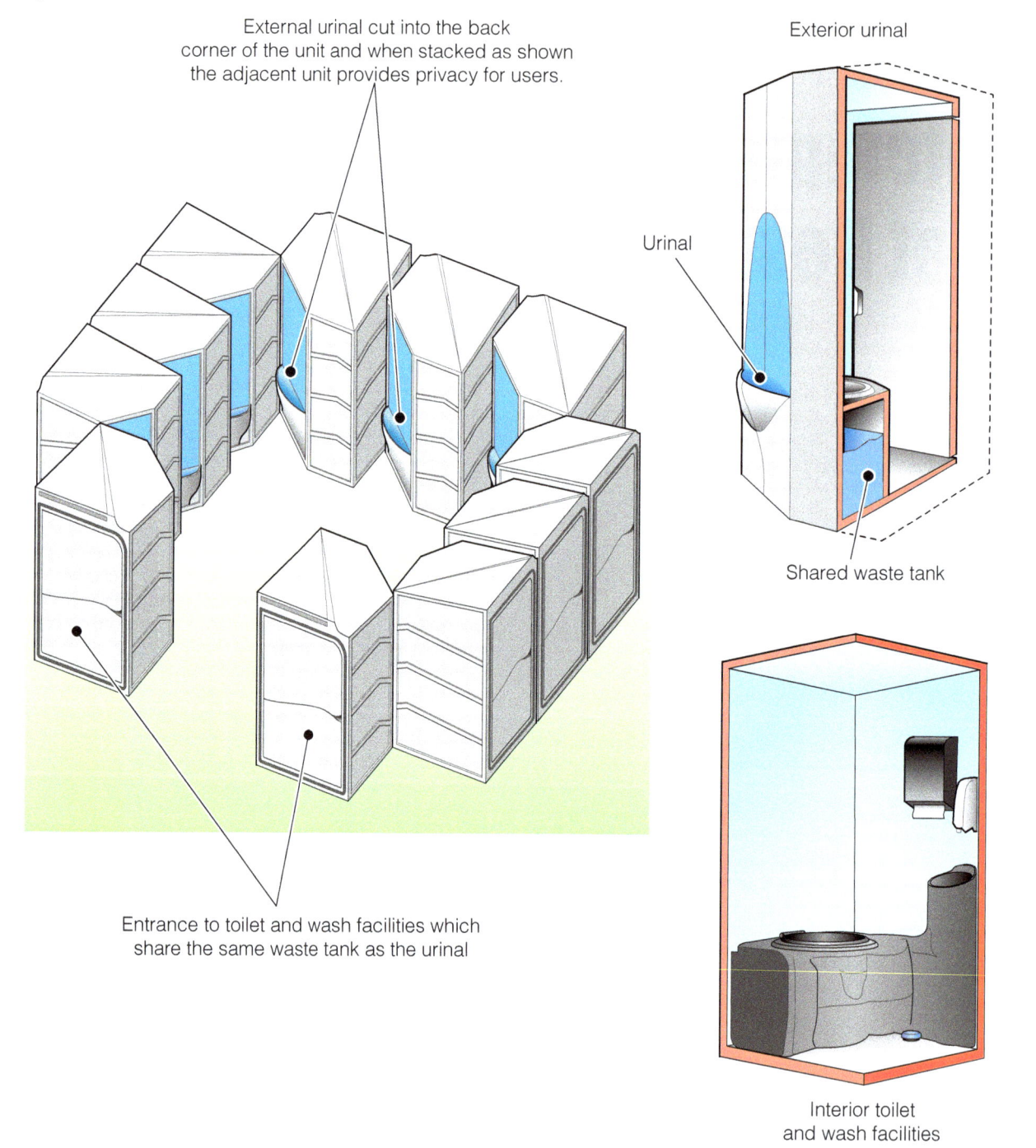

External urinal cut into the back corner of the unit and when stacked as shown the adjacent unit provides privacy for users.

Exterior urinal

Urinal

Shared waste tank

Entrance to toilet and wash facilities which share the same waste tank as the urinal

Interior toilet and wash facilities

Links
http://www.core77designawards.com/2011/recipients/2p-portable-restroom/
http://www.core77designawards.com/wp-content/uploads/2011/07/Products-Student-e588-d.pdf

Aquatron Toilet

A composting water saving toilet system separating solids from fluids for composting and wastewater treatment. The Aquatron can flush 3 litres at a time. The separation of liquids from effluent significantly improves efficiency and the unit can operate up to 20 metres from the toilet making it easy to use and relatively inexpensive.

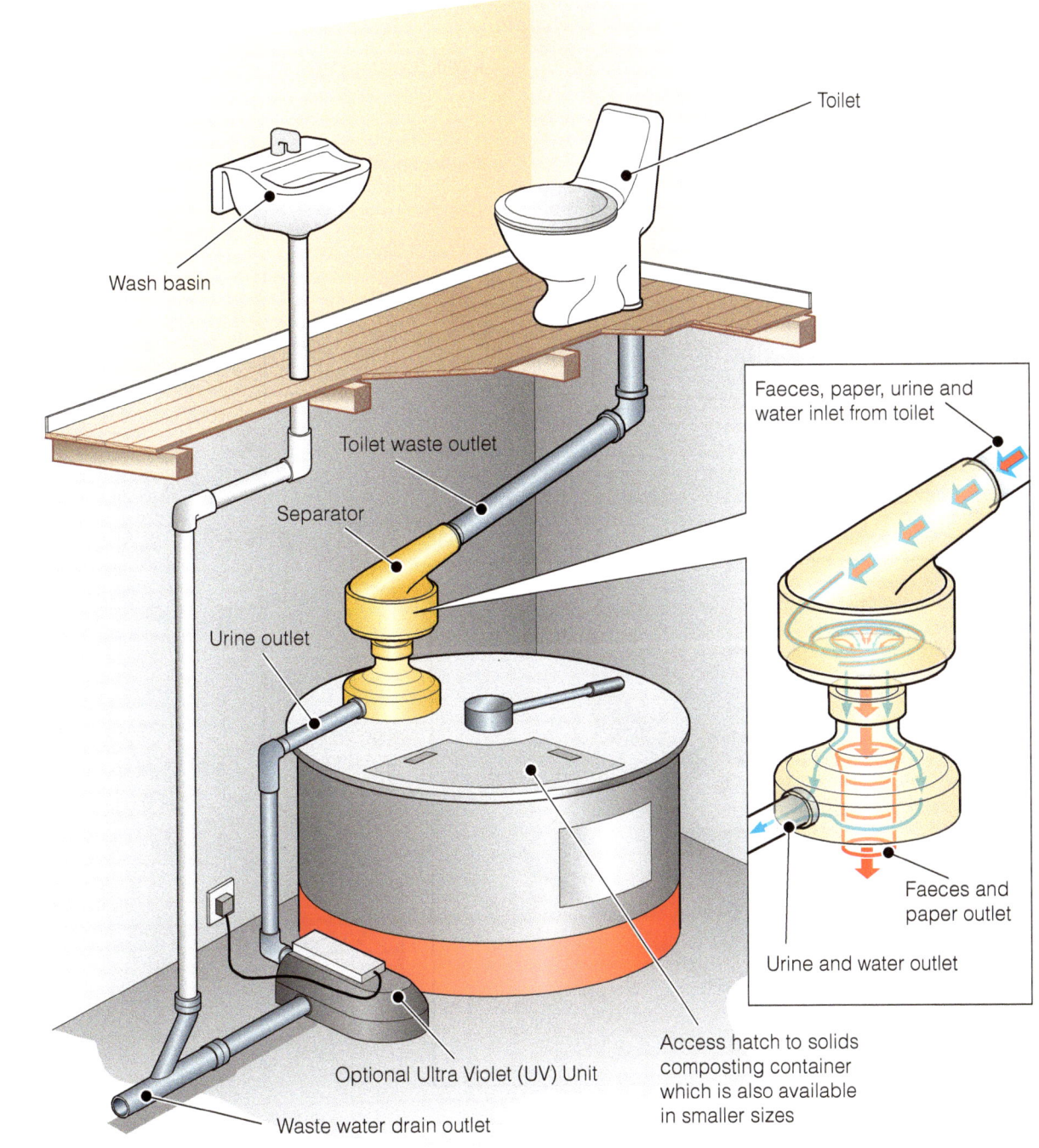

Links
http://www.naturbauhof.de/lad_komp_aqua.php

Blue Diversion Toilet

Contemporary Toilet Designs

The Blue Diversion Toilet is a dry urine-diverting toilet with the additional feature of an integrated water recovery/ recycling that allows for the comfort of hand washing, anal cleansing and toilet cleaning. Refer to the link below for full details.

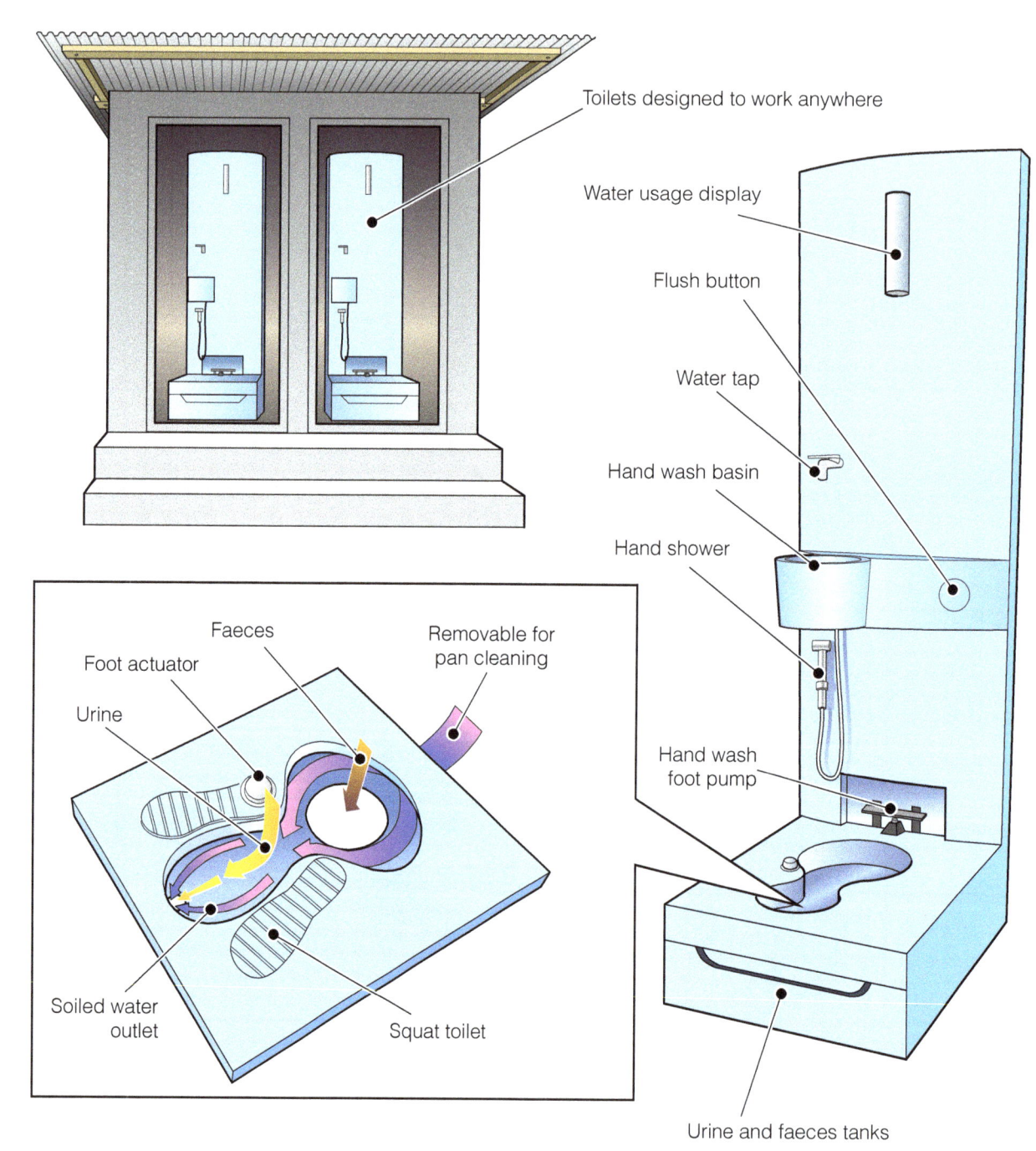

Links
http://www.bluediversiontoilet.com

Bipee*

The Bipee is an integrated water-saving bidét and urinal toilet. With the separation of urine the Bipee requires less than 0.5 litres for flushing against other conventional water-saving toilets which require more than 2.5 litres per flush.

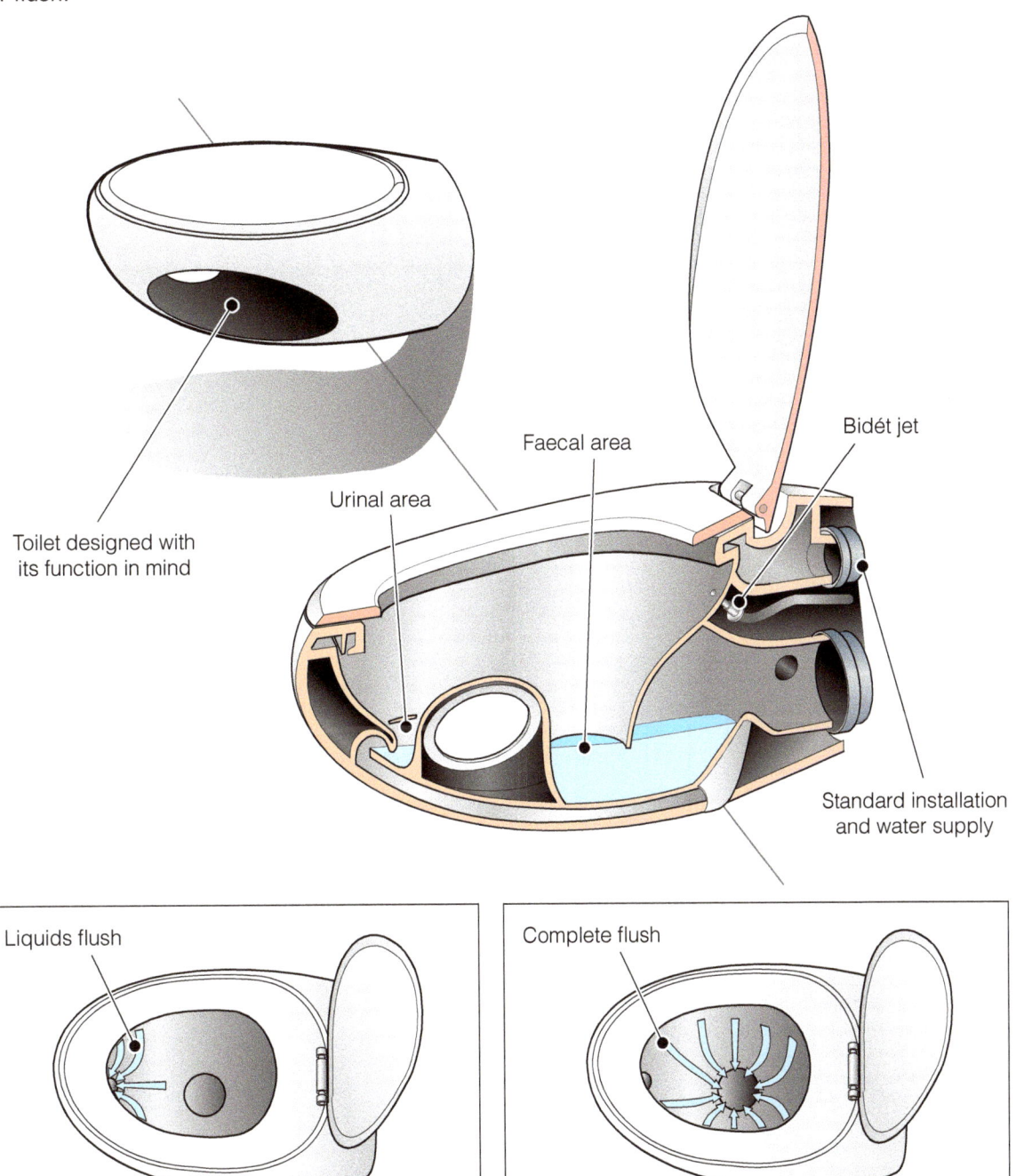

Links
http://www.m-attia.com/2010.html

Contemporary Toilet Designs

Built-in Composting System

Designed by Amos Bender, this composting system has an indoor commode, with an outside soil bin, and outside access to a waste container. The system incorporates a small amount of soil mixed with the faeces to assist with composting. (See also pages 19 and 28.)

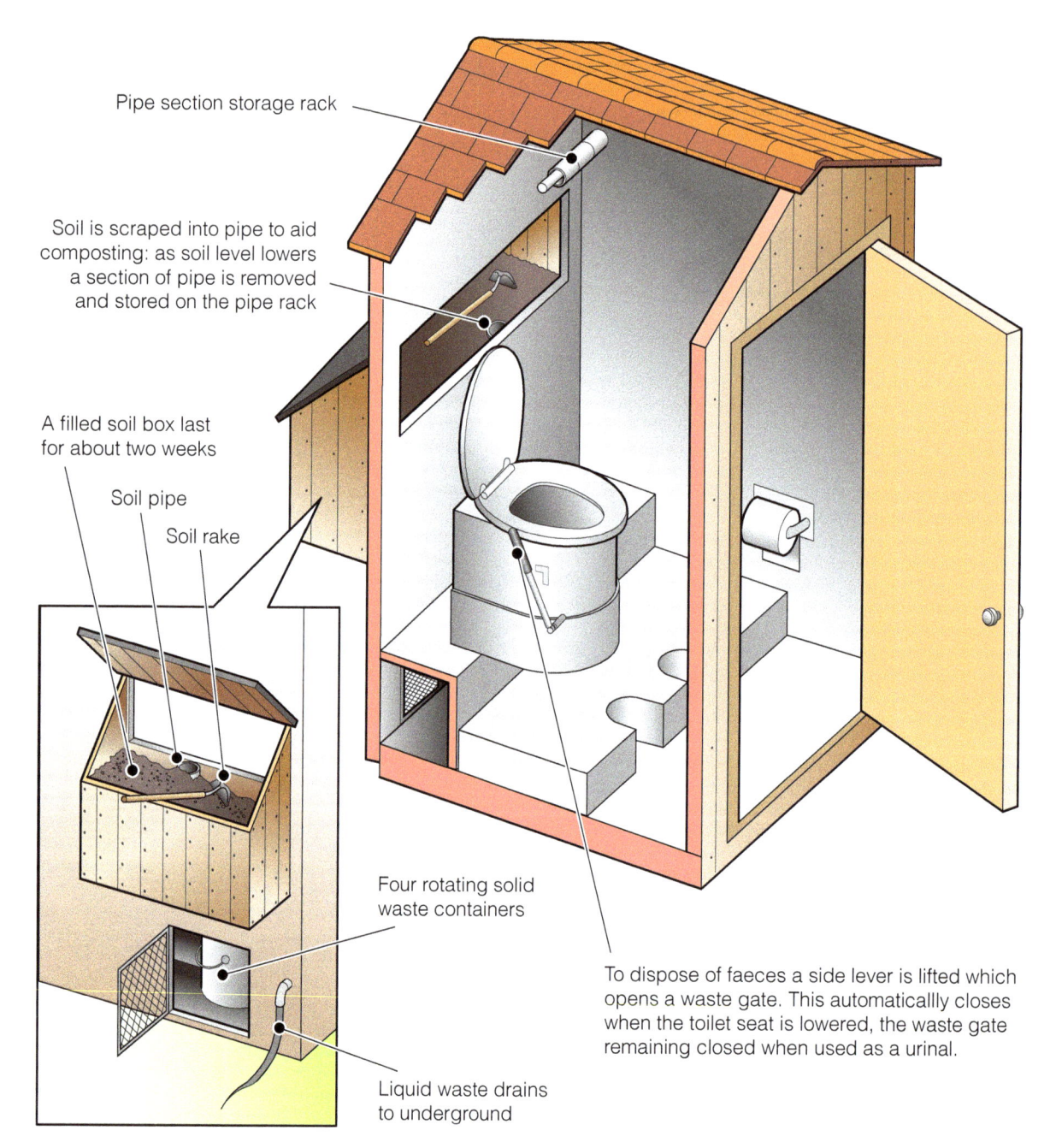

Pipe section storage rack

Soil is scraped into pipe to aid composting: as soil level lowers a section of pipe is removed and stored on the pipe rack

A filled soil box last for about two weeks

Soil pipe

Soil rake

Four rotating solid waste containers

Liquid waste drains to underground

To dispose of faeces a side lever is lifted which opens a waste gate. This automaticallly closes when the toilet seat is lowered, the waste gate remaining closed when used as a urinal.

Links
http://www.exploringnaturespossibilities.com/

Camping Toilets

Contemporary Toilet Designs

Camping toilets are portable and are produced by various manufacturers. They usually use chemical solutions as a primary means of treatment at the point of use. A world-wide-web search query produces a comprehensive list of links.

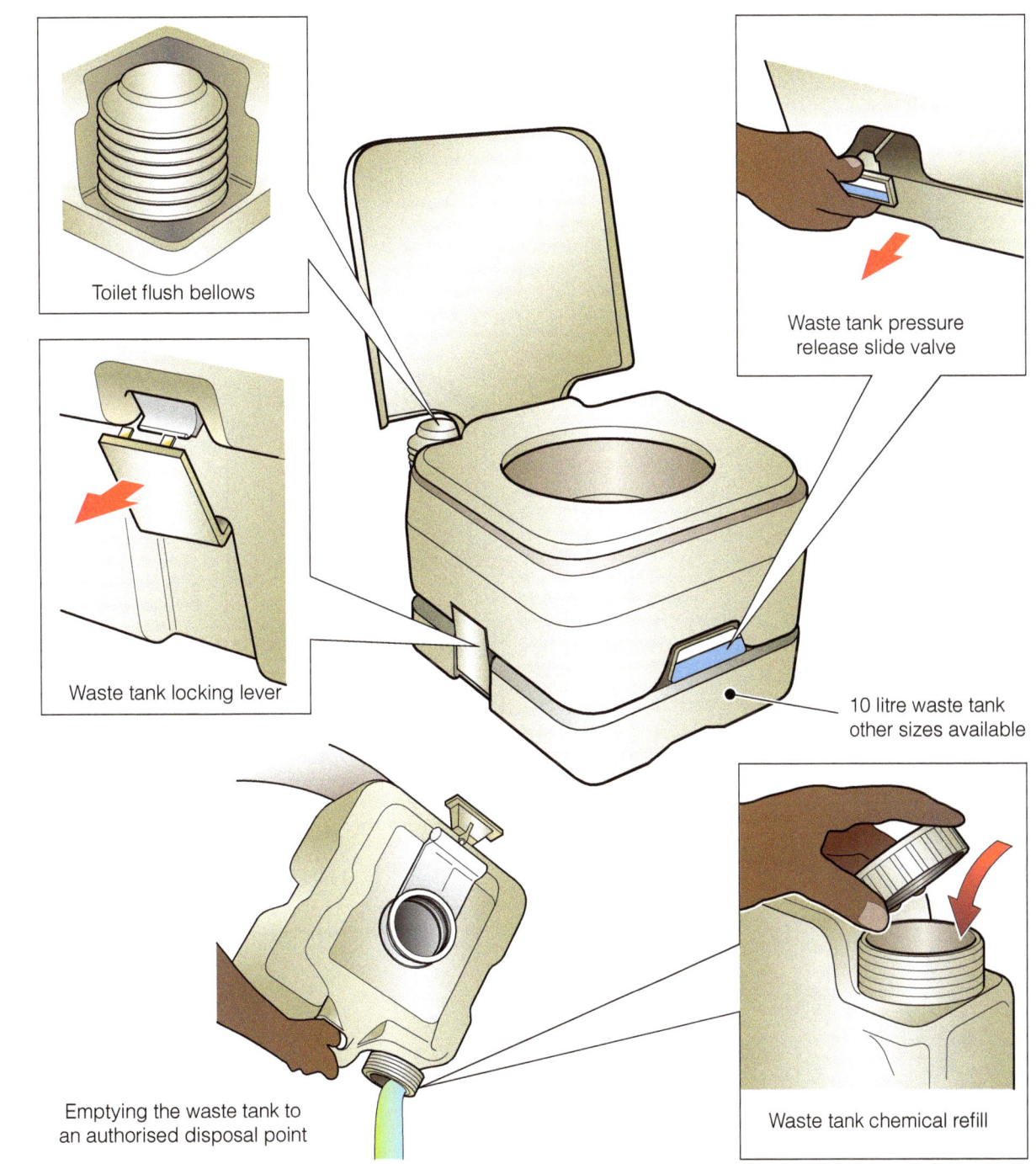

Links

http://www.youtube.com/watch?v=vhELPhb--Lg
http://www.kampa.co.uk/instructions/accessories/portaflush

Contemporary Toilet Designs

Caravan Toilets

Caravan toilets are similar to camping toilets but are designed to be integrated into the chassis of the vehicle. They are removed for periodic emptying at chemical disposal points.

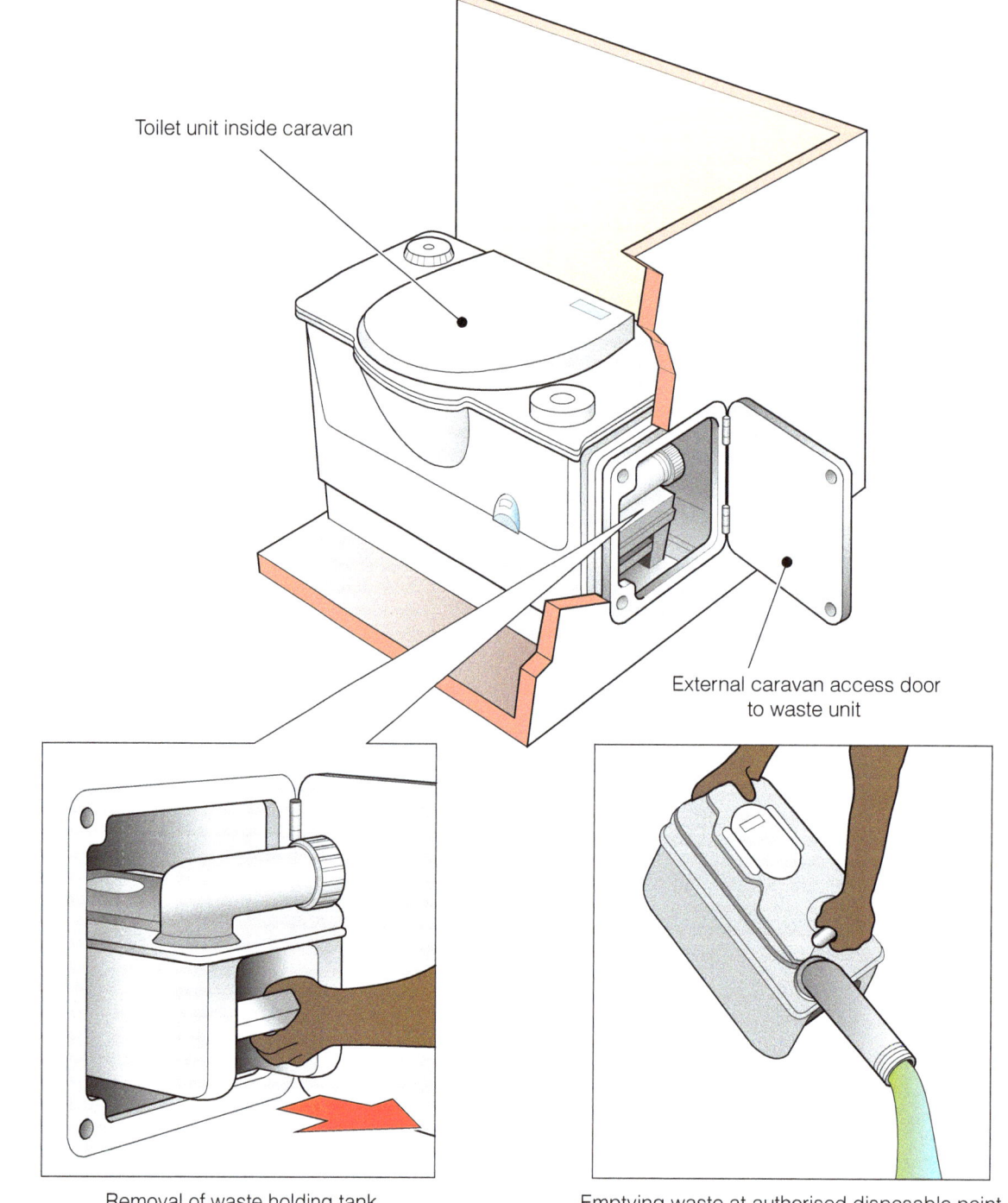

Toilet unit inside caravan

External caravan access door to waste unit

Removal of waste holding tank

Emptying waste at authorised disposable point

Links
http://www.youtube.com/watch?v=CQ5Vvjqnjjg
http://www.thetford.com

Contemporary Toilet Designs

Cinderella Toilet

Cinderella toilet systems are incinerating toilets, in which the waste products are converted into sanitized ash by combustion at high temperatures.

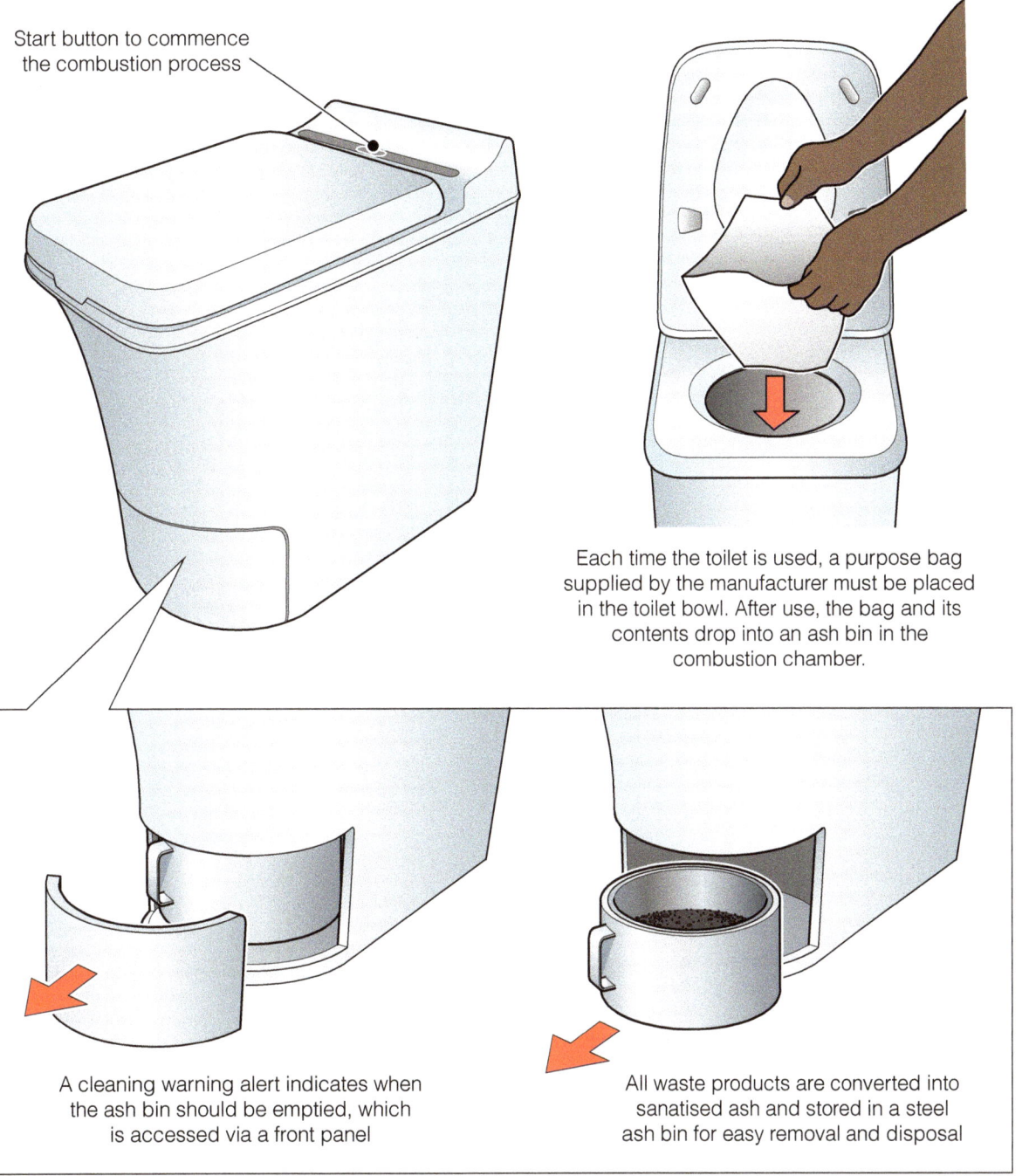

Start button to commence the combustion process

Each time the toilet is used, a purpose bag supplied by the manufacturer must be placed in the toilet bowl. After use, the bag and its contents drop into an ash bin in the combustion chamber.

A cleaning warning alert indicates when the ash bin should be emptied, which is accessed via a front panel

All waste products are converted into sanatised ash and stored in a steel ash bin for easy removal and disposal

Links
http://www.cinderella.as

Drysan Waterless Toilet

The Drysan waterless and urine diversion toilet system is intended for use in areas where sewerage lines and sewerage works are unavailable, have been damaged, or where chemical toilets are currently used. The system may also be used where households have a limited supply of water. The use of the Drysan extends to informal housing, low-cost housing, smallholdings, disaster areas, water challenged areas, military facilities, lodges, farms and many other areas deprived of water toilet facilities.

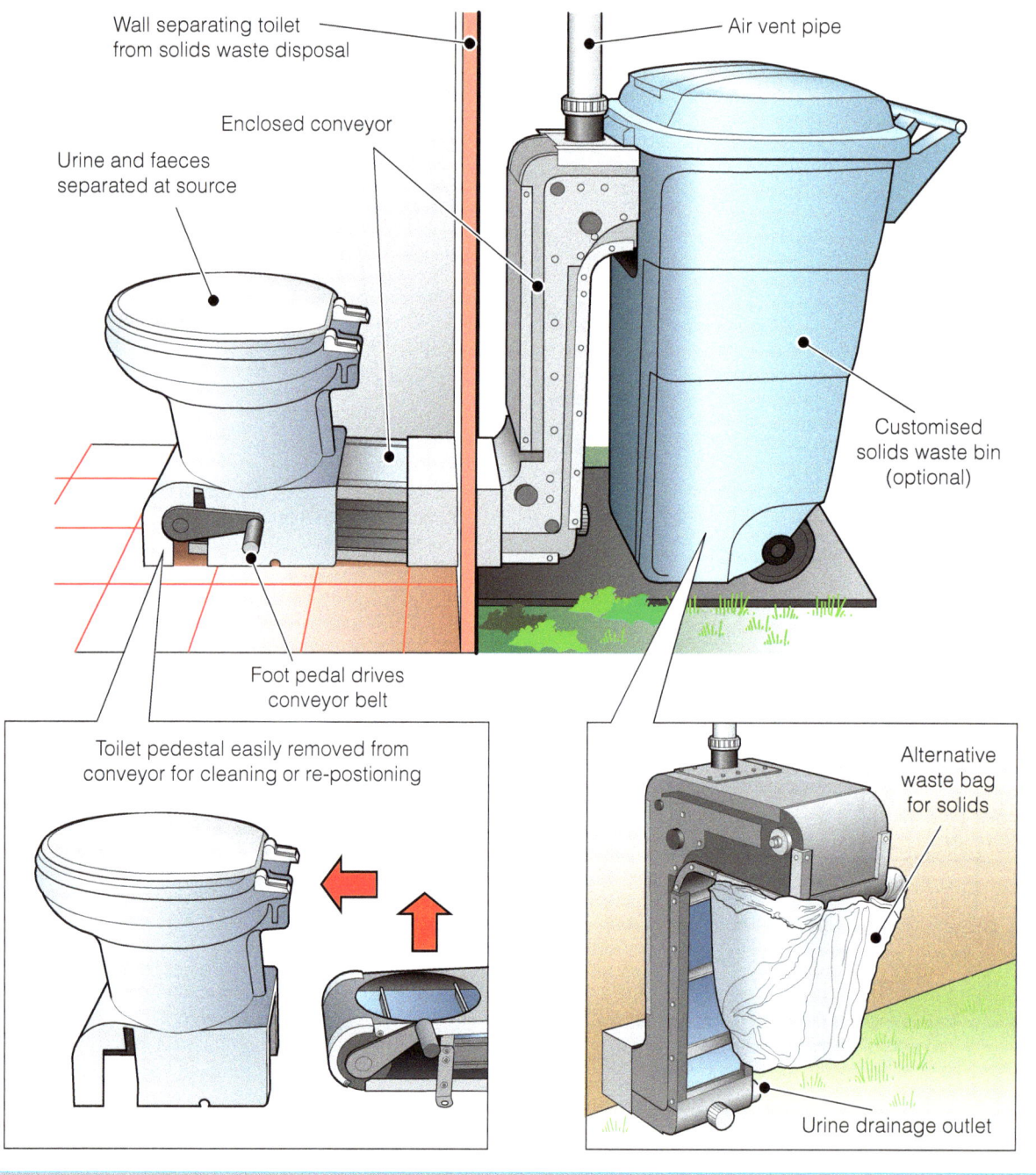

Links
http://drysan.co.za/

Contemporary Toilet Designs

EkoToi

Designed for both rural and urban areas in India, the EkoToi is the result of a Masters final project in product design from Parag Deshpande, featuring urine separation.

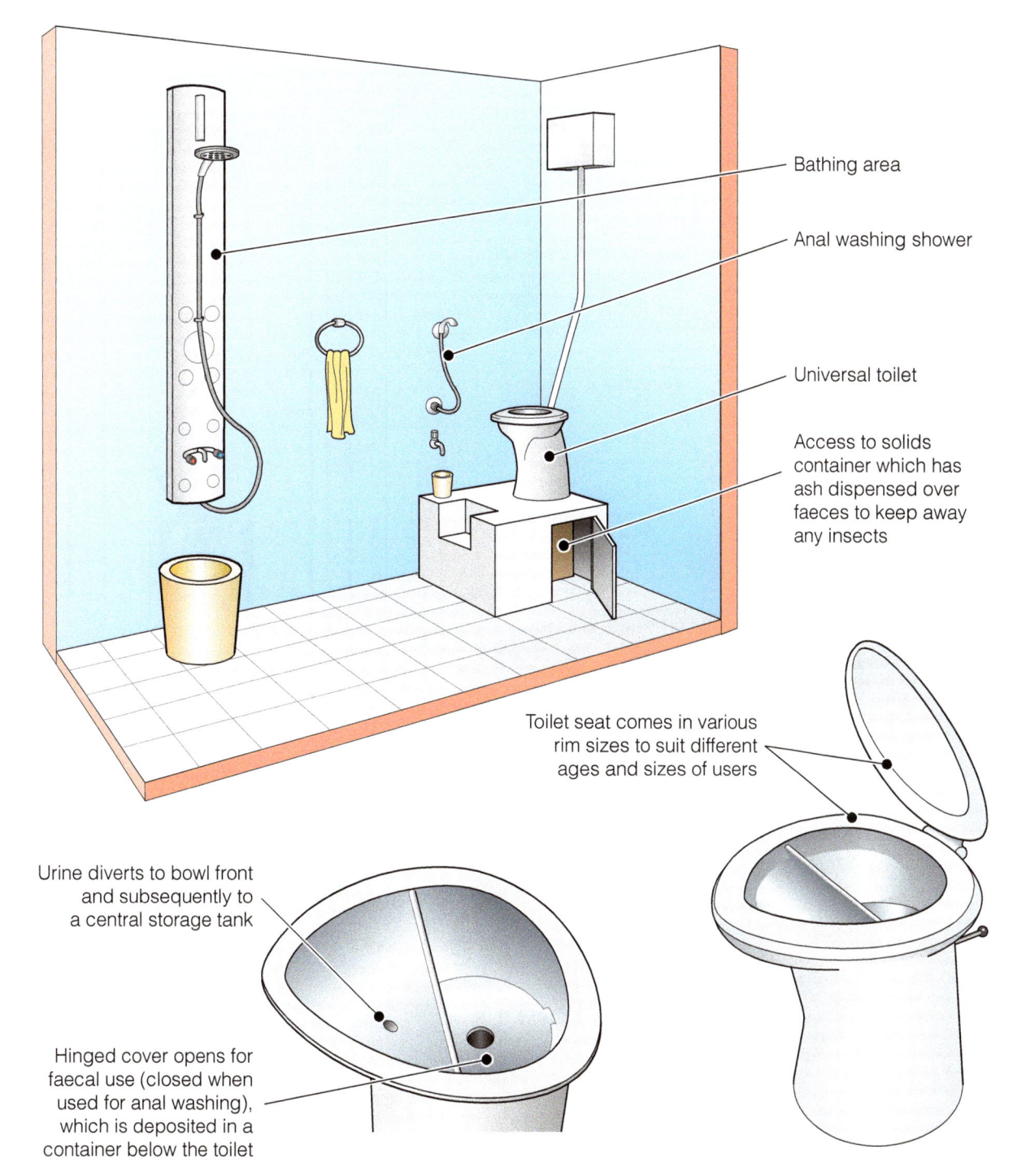

- Bathing area
- Anal washing shower
- Universal toilet
- Access to solids container which has ash dispensed over faeces to keep away any insects
- Toilet seat comes in various rim sizes to suit different ages and sizes of users
- Urine diverts to bowl front and subsequently to a central storage tank
- Hinged cover opens for faecal use (closed when used for anal washing), which is deposited in a container below the toilet

Links
http://vimeo.com/user2830873

Envirosan

Based in South Africa, Envirosan provides a comprehensive range of plastic injection moulding products, concentrating specifically on environmentally-friendly sanitation systems.

The VIP480 injection moulded pedestal, including an intergrated flange fitting into a cast-in-support ring, reducing movement of the pedestal and so increasing safety

Links
http://envirosan.co.za/

Contemporary Toilet Designs

Eram Delight

Delight, the automatic public toilet unit developed by Eram Scientific Solutions Pvt. Ltd is intended to improve the sanitation facilities in urban areas, especially in developing countries. In addition to its automated toilet facilities the unit can also display 100sq ft of illuminated advertisement boards on its outer surface. This can act as a valuable source of advertisement revenue in addition to the admission collection for those operating the toilet. The unit can be connected to existing water and drainage facilities or a bio-membrane reactor which can recycle the wastewater for flushing purposes.

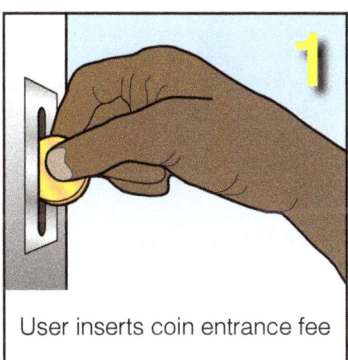
User inserts coin entrance fee

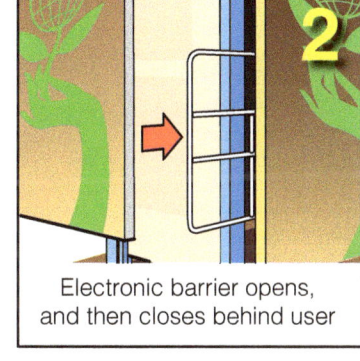
Electronic barrier opens, and then closes behind user

Toilet is automatically cleaned and flushed after each user

User presses a button when finished, to open barrier

Toilet operator can access toilet data either by computer...

...or by smartphone

Links

http://www.susana.org/docs_ccbk/susana_download/2-1624-baby.pdf
http://www.youtube.com/watch?v=1bBhO7AQZao
http://www.youtube.com/watch?v=IYsygg1ZJAg

Contemporary Toilet Designs

Fresh Life Toilet

Sanergy design and manufacture low-cost, high-quality sanitation facilities. Developed by their engineers, the Fresh Life Toilet (FLT) is pre-fabricated at their local workshop. The FLT qualities users value most include the high-quality materials that are easy to keep clean and maintain; the small footprint that enables them to be installed close to homes; and essential features such as hand washing facilities.

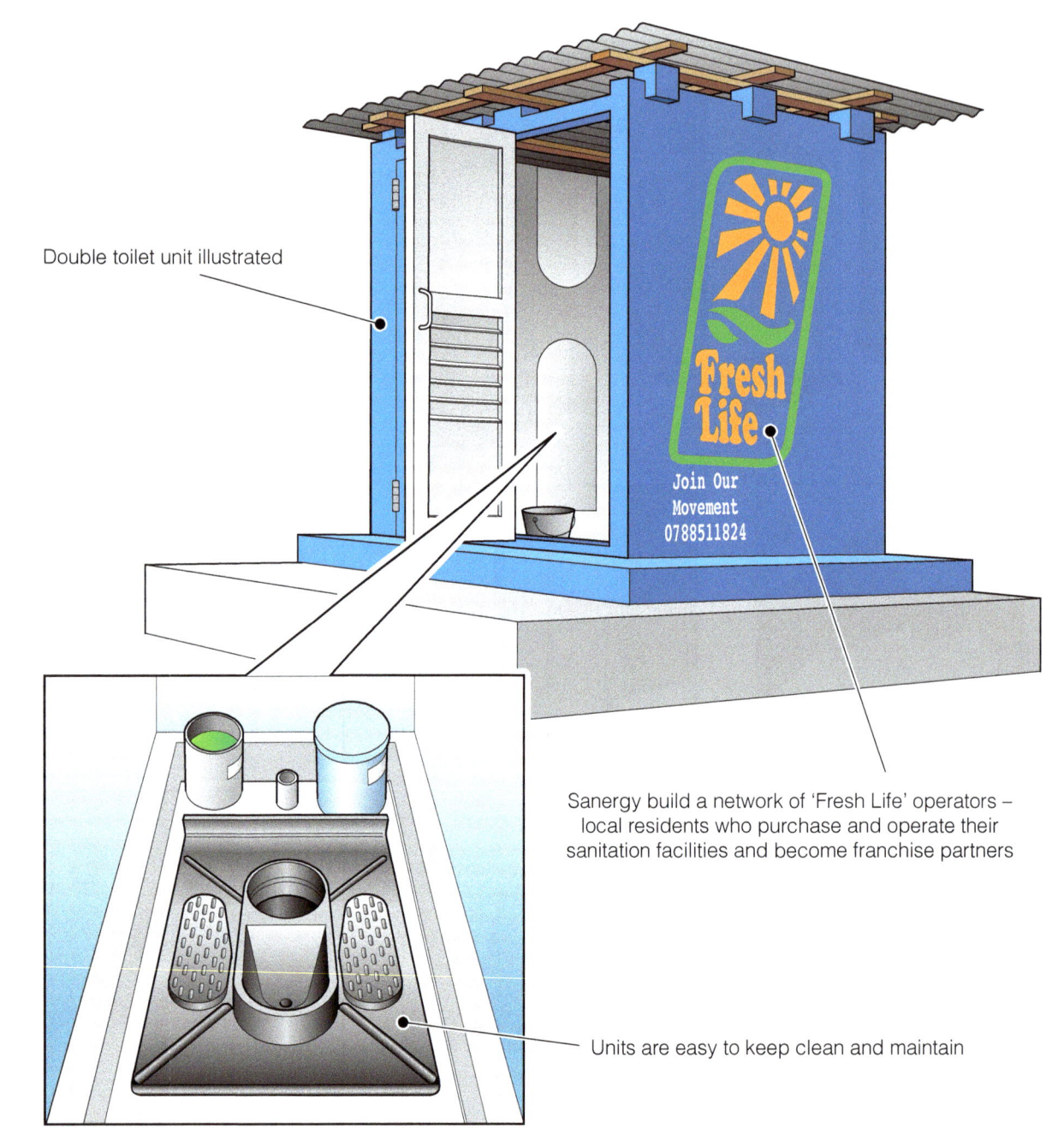

Double toilet unit illustrated

Sanergy build a network of 'Fresh Life' operators – local residents who purchase and operate their sanitation facilities and become franchise partners

Units are easy to keep clean and maintain

Links
http://saner.gy/
http://www.youtube.com/watch?v=unTc-rID9LI
http://www.flickr.com/photos/gtzecosan/sets/72157629321008423/with/6879710585/

14

Contemporary Toilet Designs

Ghanasan

The Ghanasan Human Centered Design Research Project is a collaboration between Unilever, Water and Sanitation for the Urban Poor (WSUP), and IDEO to develop new products and services for in-home sanitation in Kumasi, Ghana.

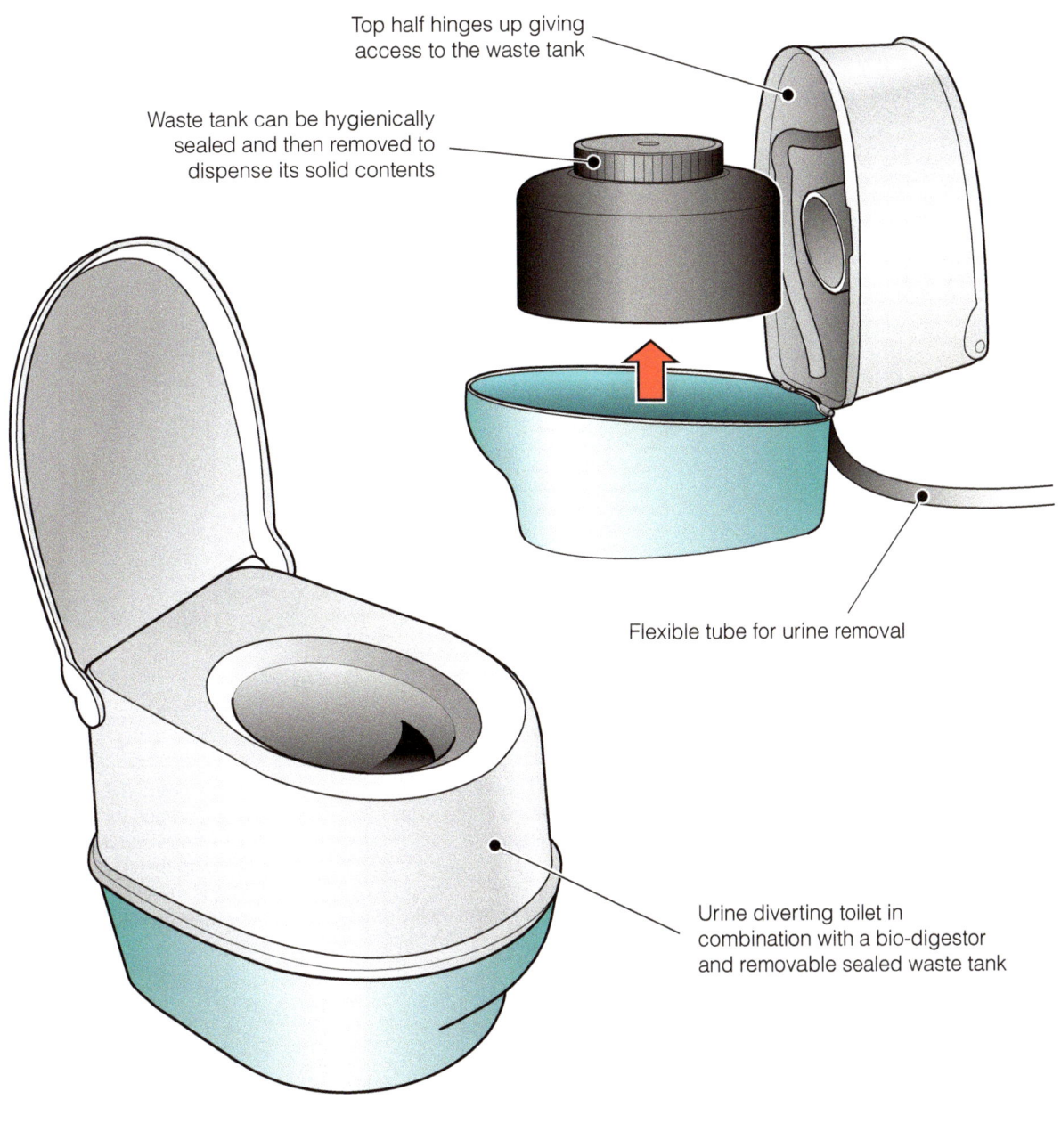

Top half hinges up giving access to the waste tank

Waste tank can be hygienically sealed and then removed to dispense its solid contents

Flexible tube for urine removal

Urine diverting toilet in combination with a bio-digestor and removable sealed waste tank

Links
http://ghanasan.wordpress.com/
http://www.ideo.com/work/human-centered-design-toolkit/

Contemporary Toilet Designs

Green, Portable Eco-toilet*

The eco-toilet provides a clean, portable sanitation solution for refugee camps. The waste solution can be combined with mobile shelters. Made of roto-moulded biodegradable plastic and ceramic, the toilet bowl is designed to provide comfort.

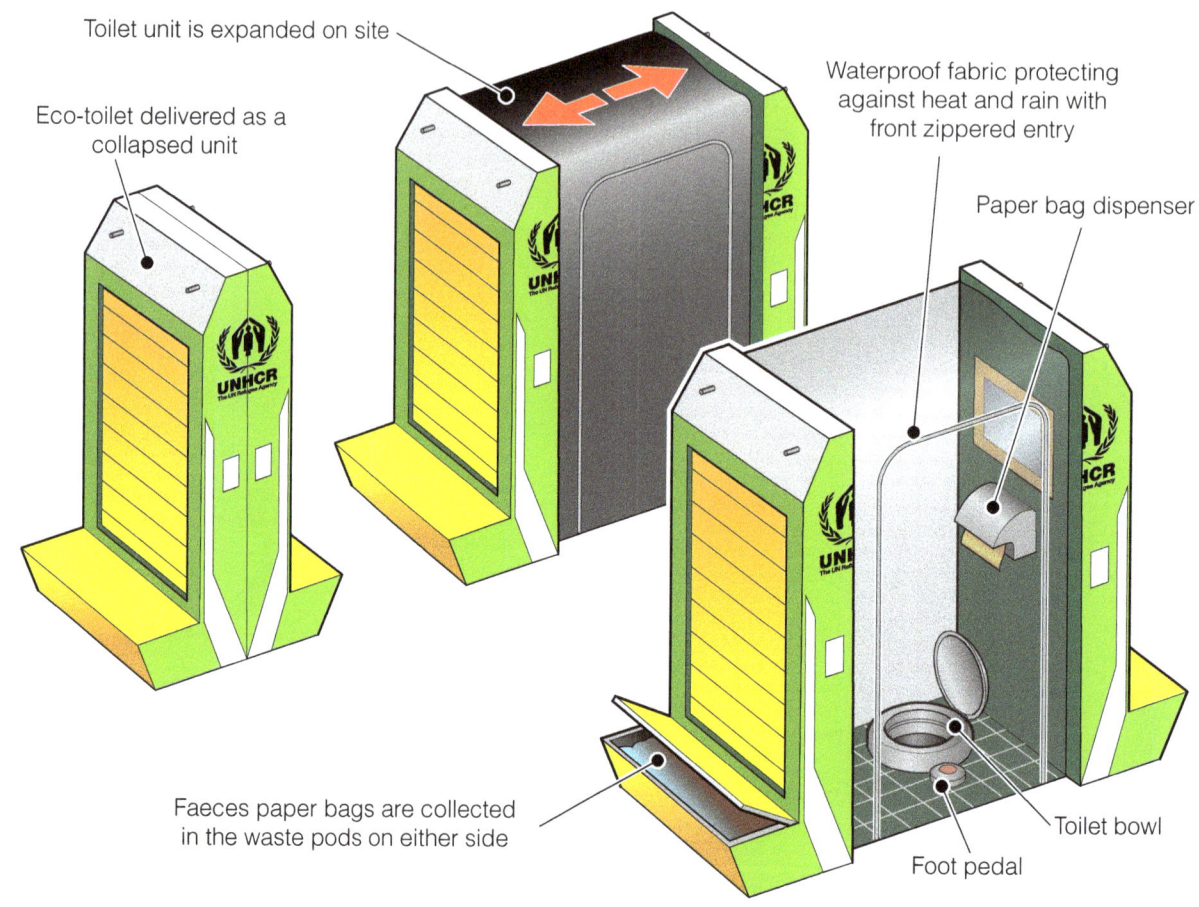

- Eco-toilet delivered as a collapsed unit
- Toilet unit is expanded on site
- Waterproof fabric protecting against heat and rain with front zippered entry
- Paper bag dispenser
- Toilet bowl
- Foot pedal
- Faeces paper bags are collected in the waste pods on either side

1. User steps on pedal to open cover like a waste bin

2. Paper bag (containing powder) is placed in the toilet bowl and seat closed

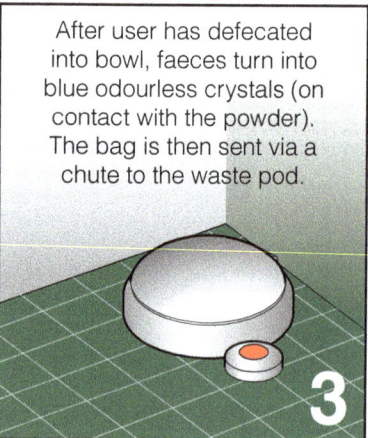

3. After user has defecated into bowl, faeces turn into blue odourless crystals (on contact with the powder). The bag is then sent via a chute to the waste pod.

Links
http://www.igreenspot.com/green-portable-toilet-an-eco-friendly-waste-disposal-area/

Contemporary Toilet Designs

Hightech Composting Toilet: Ecodomeo

Ecodomeo dry toilets separate the liquids from the solids. Liquids are evacuated through the household used water system. Solids are evacuated to a closed space where they are then reduced to compost by earthworms.

- Air vent pipe
- Conveyor deposits solids (faeces and toilet paper) material into a suitable composting space which is then further treated by earthworms
- Note: Urine is sent to a tank within the toilet unit which in turn delivers it to an evacuation pipe and the sewage system
- Removable internal toilet seat housing
- External earthworm composting unit
- Conveyor cover
- Conveyor belt outlet
- Foot pedal which deposits all solid waste to the conveyor belt
- Conveyor base screwed to floor

Toilet housings come in alternative designs

Links
http://www.ecodomeo.com/english/

17

IDE EZ Latrine

The Sanitation Marketing Pilot Project carried out by IDE Cambodia's WATSAN team is designed with the overall objective of creating rural household demand for sanitation, and linking this demand to local suppliers who have been educated, trained and supported to deliver sustainable, low-cost latrines. The end result will be a continually developing and thriving private marketplace that satisfactorily addresses the need for rural sanitation.

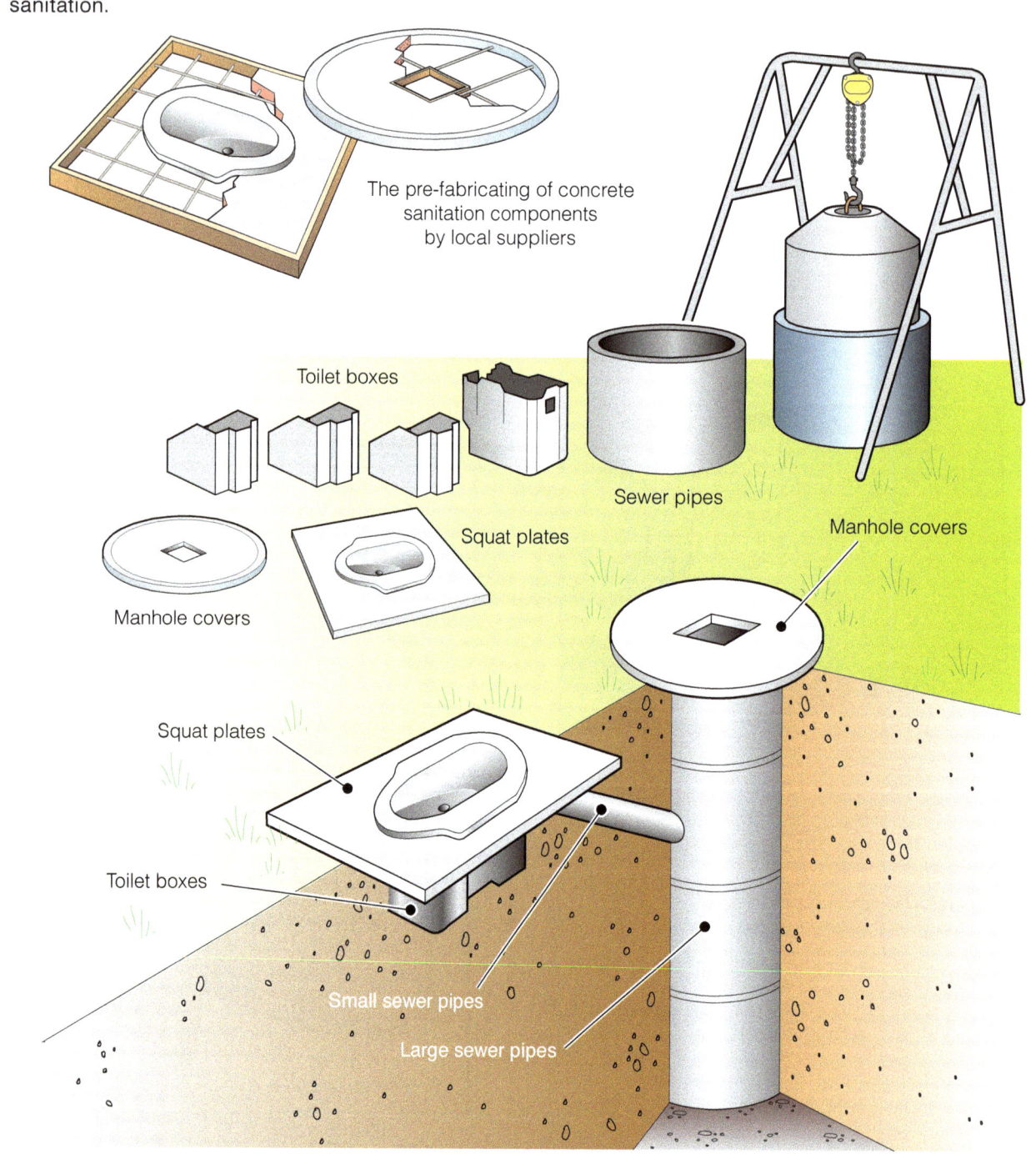

Links
http://www.youtube.com/watch?v=zIoOePlhQzc

Indoor Self-contained Toilet

Contemporary Toilet Designs

Another system from Amos Bender, this is a soiless, largely waterless, self contained, indoor commode system for use in power outages, cabins, or simply for use in homes. (See also Pages 6 and 28).

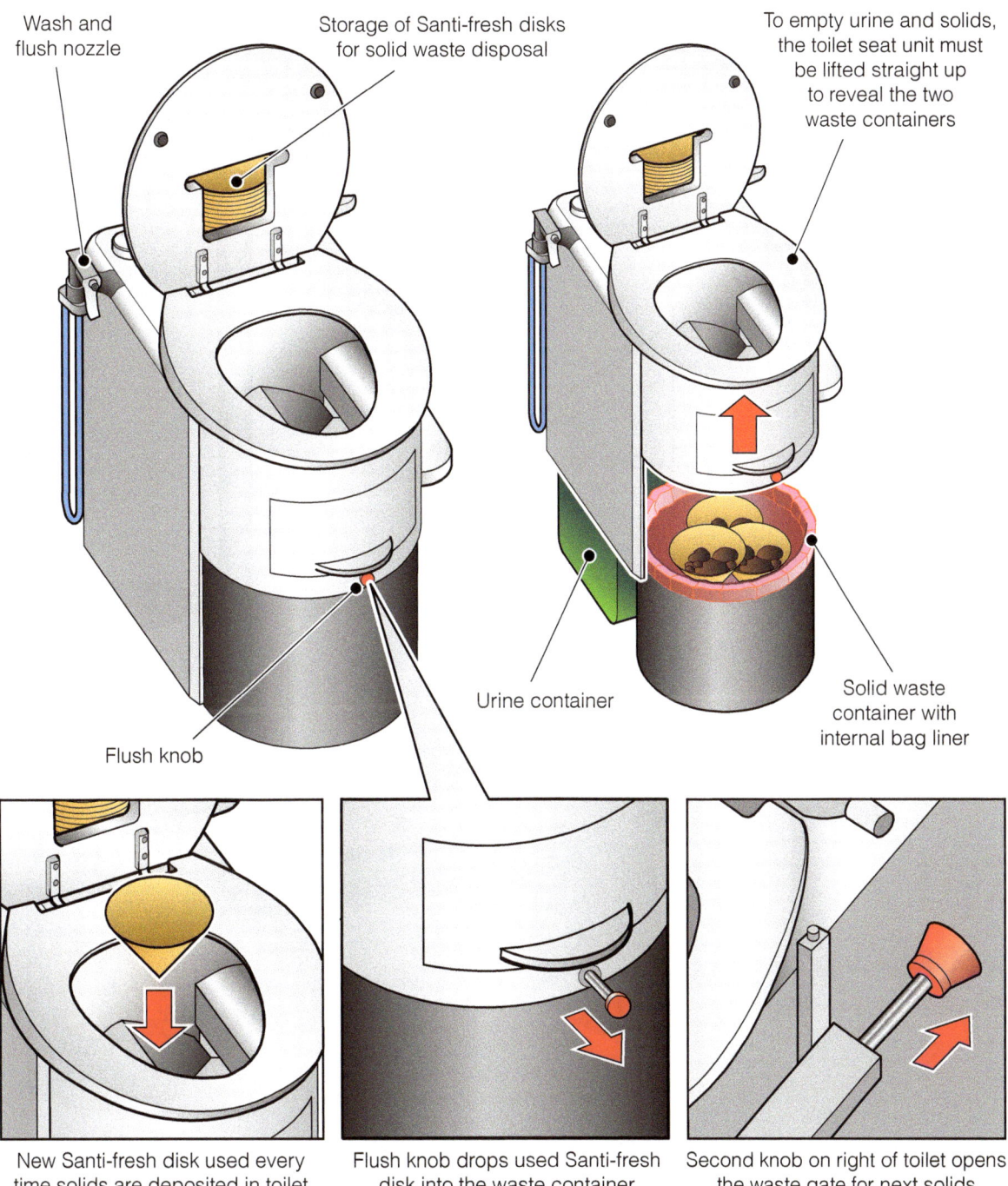

- Wash and flush nozzle
- Storage of Santi-fresh disks for solid waste disposal
- To empty urine and solids, the toilet seat unit must be lifted straight up to reveal the two waste containers
- Flush knob
- Urine container
- Solid waste container with internal bag liner
- New Santi-fresh disk used every time solids are deposited in toilet
- Flush knob drops used Santi-fresh disk into the waste container
- Second knob on right of toilet opens the waste gate for next solids

Links

http://www.exploringnaturespossibilities.com/

Intestinal Toilet

With the intestinal toilet, excreta falls down a vertical chute onto one end of a specially designed helical screw conveyor. Every time the toilet lid is lifted, a mechanism rotates the conveyor. With each rotation the excreta slowly moves along, taking approximately twenty five days before falling into a reusable collection bag. It takes six months for the bag to fill with dry and odourless waste. Through the ventilation pipe, adequate airflow is provided for the dehydration, evaporation and deodorising process.

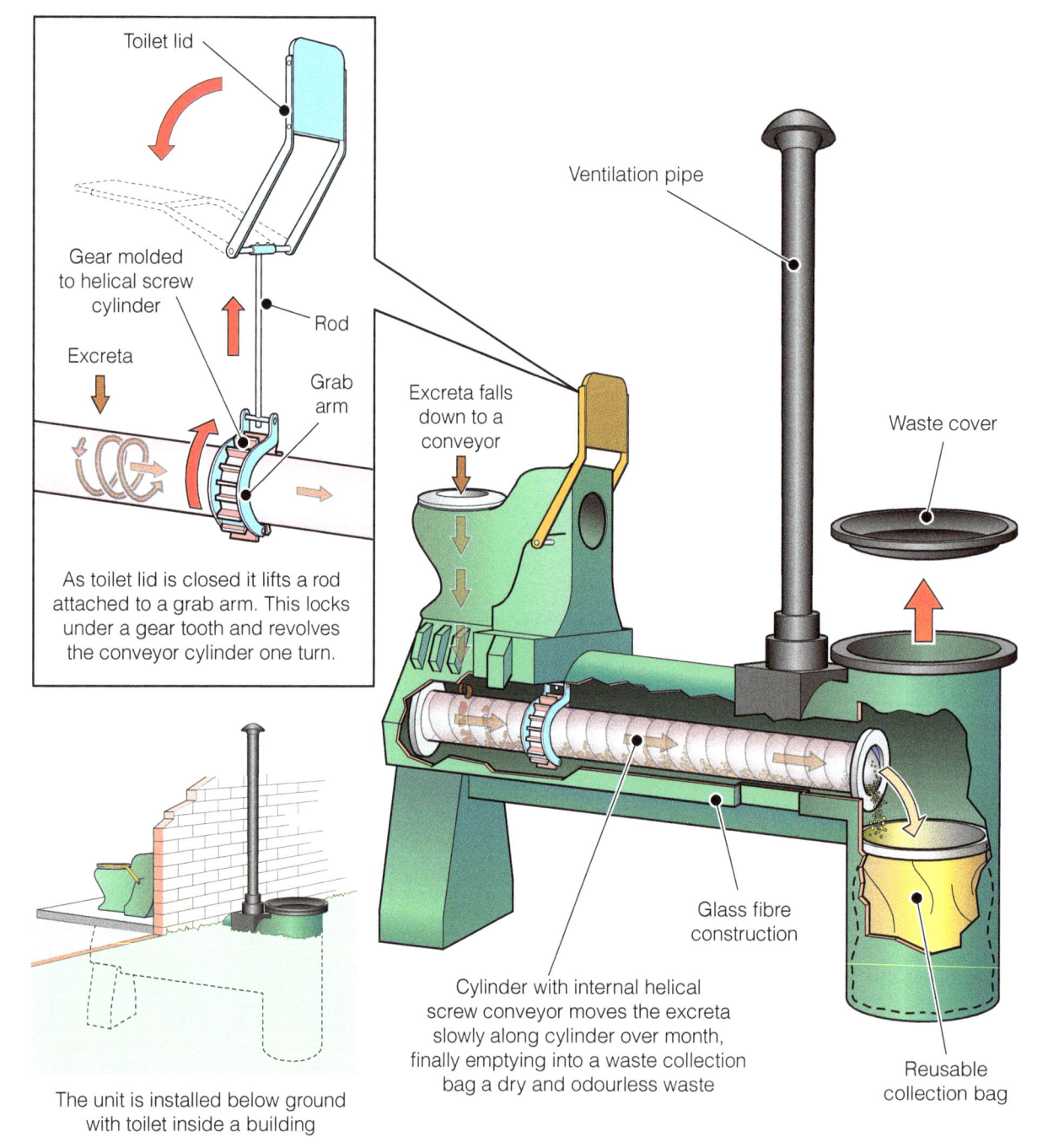

As toilet lid is closed it lifts a rod attached to a grab arm. This locks under a gear tooth and revolves the conveyor cylinder one turn.

The unit is installed below ground with toilet inside a building

Cylinder with internal helical screw conveyor moves the excreta slowly along cylinder over month, finally emptying into a waste collection bag a dry and odourless waste

Links
http://www.youtube.com/watch?v=gPummZRR2Cg
http://www.ecosan.co.za/

Locus Toilet

Contemporary Toilet Designs

The Locus toilet is a biological system for waste treatment in which excess liquid evaporates and the faeces are composted by natural microorganisms. A controlled supply of heat and air and frequent stirring accelerates the decomposition of the waste, producing a valuable fertilizer.

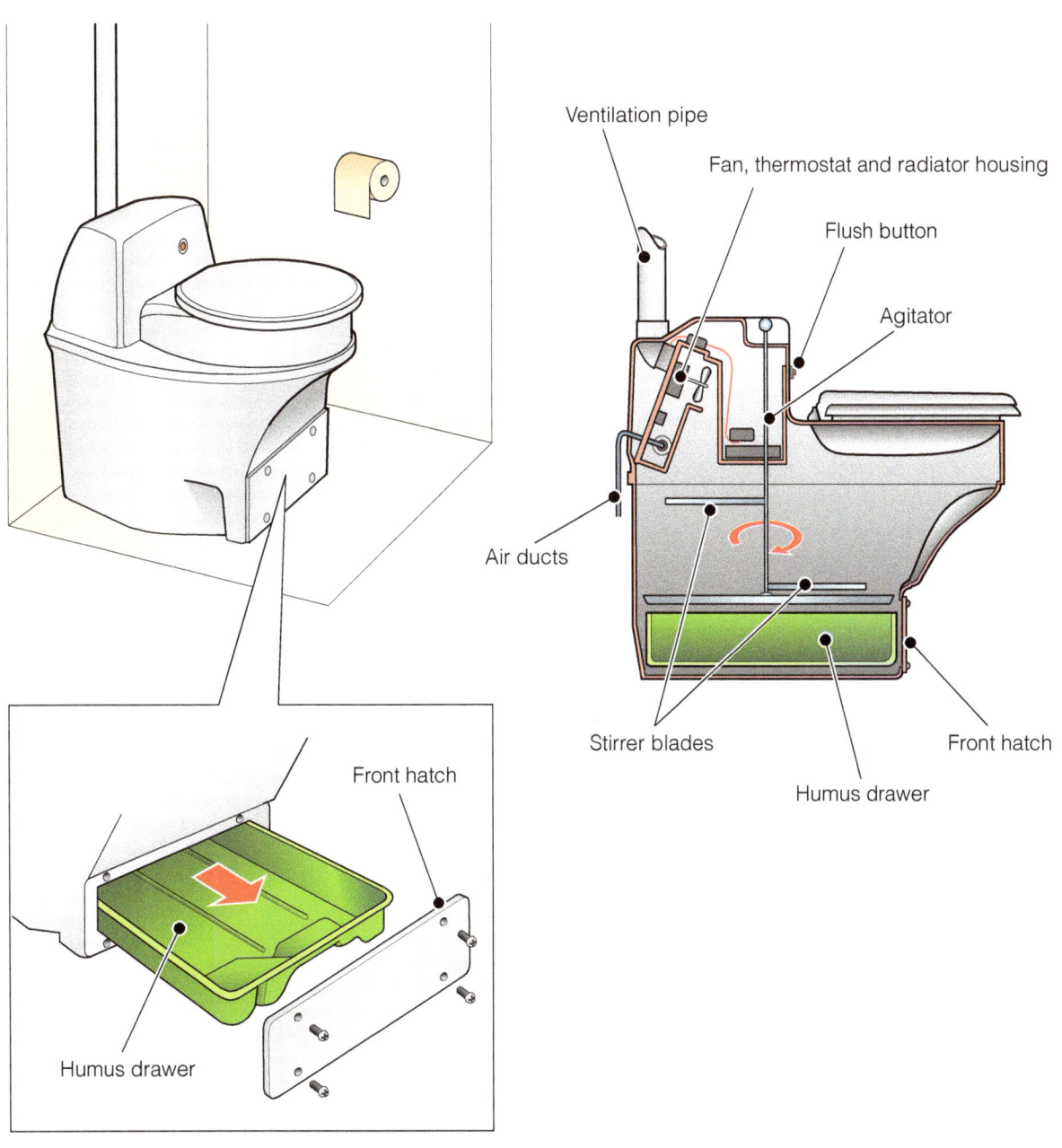

Links
http://www.locus-toilette.de/umwelt/umwelt.htm

Loowatt Toilet

Contemporary Toilet Designs

The Loowatt toilet uses a simple, patent-protected mechanical sealing unit to contain human waste within biodegradable film in the most efficient way possible, with a unique odour-inhibiting system. The waste is then stored in a cartridge beneath the toilet, for periodic emptying, which can be weekly or daily, depending upon level of usage and capacity. The sealing unit can be built into toilets of any shape, size and specification, using off-the-shelf parts and local materials to maximize value. In addition, they also market a biodegradable digester for extracting methane gas from their cartridges for use in cooking.

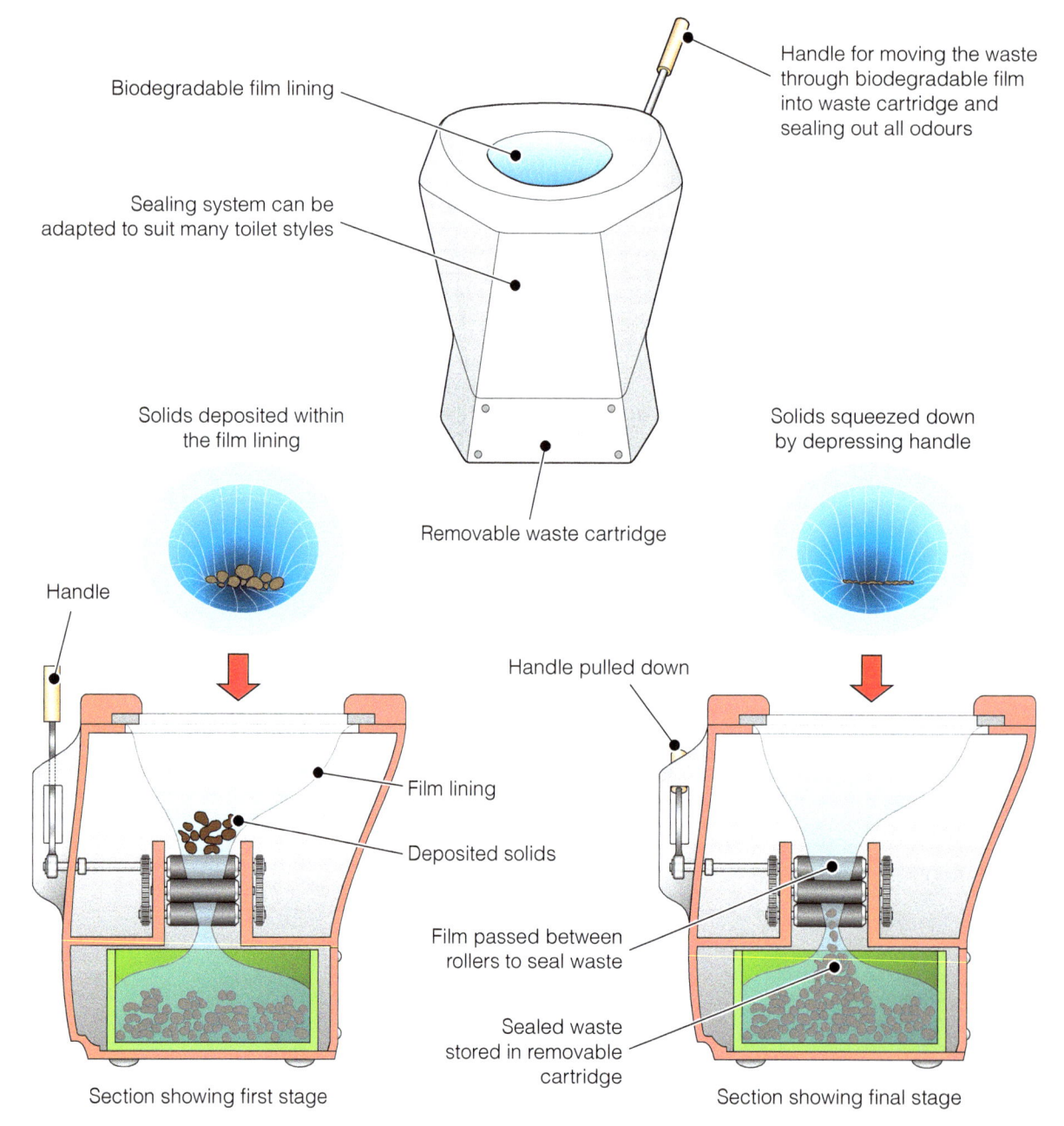

Links
http://www.loowatt.com

MoSan – Mobile Sanitation

GIZ Bangladesh and the german industrial designer Mona Mijthab developed a non infractructure-based separation dry toilet for the urban poor in Bangladesh. Called the MoSan, the toilet is part of a sustainable sanitation system with collectors, transport, human waste treatment and reuse as compost fertiliser or biogas. No water or chemicals are required. The lightweight toilet can be carried and used at a convenient place and is also suitable in cases of emergency, like flooded areas. The toilet is designed for one family or household.

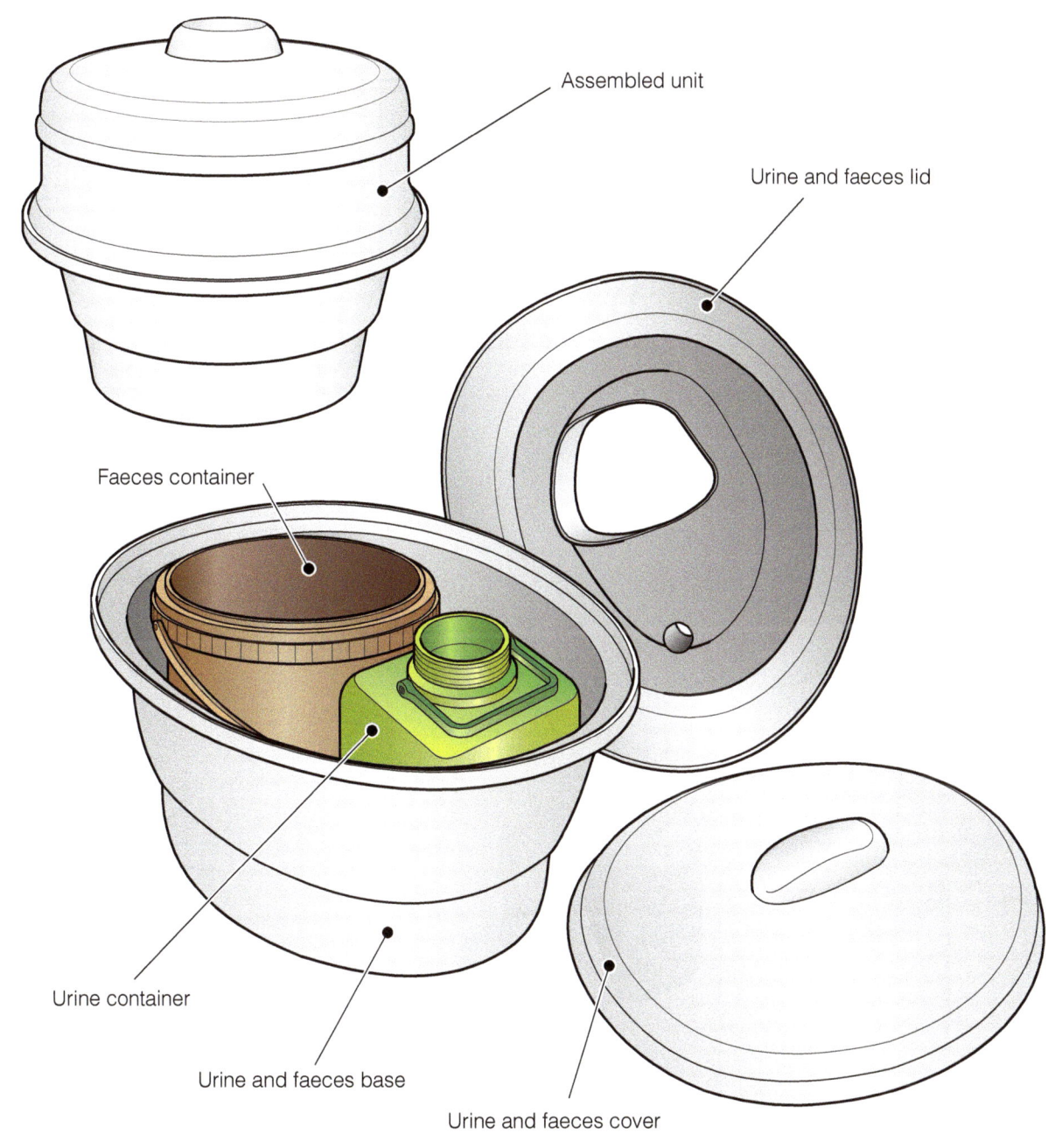

Links
http://mosan-bangladesh.tumblr.com/
http://www.susana.org/lang-en/library?view=ccbktypeitem&type=2&id=1237

Contemporary Toilet Designs

Open Privy*

A waterless composting toilet, the Open Privy can be installed outdoors or within a superstructure. If it is placed in the open, rainwater can be collected and allowed to filter down to a downpipe. Bent metal and willow sticks have been woven to cover the toilet to guard against weathering. A hinge that connects the toilet's inlay, seat, downpipe and cover masks the smell, letting air circulate freely.

The Open Privy serves as a good method of composting and enhances the top soil effectively to act as a natural fertilizer. Users can choose to either throw away the excreta collected or reuse it in the composting process.

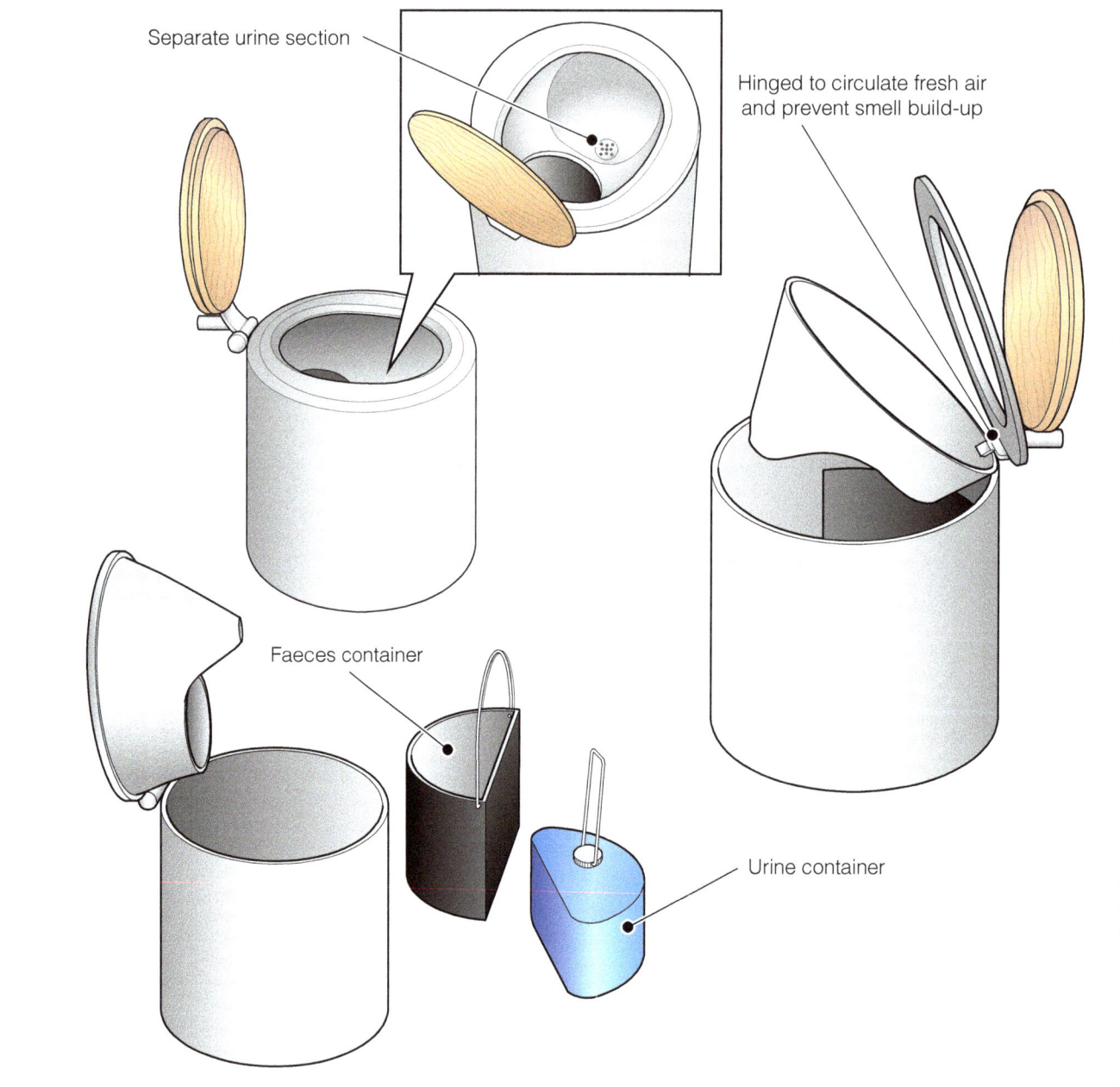

Links
http://www.youtube.com/watch?v=1LvsvLlnkN8
http://www.ecofriend.com/open-privy-waterless-toilet-helps-add-precious-top-soil.html

Otji Toilet

The Otji Toilet from Namibia uses the effect of surface tension to divert faeces and urine. The superstructure typically used is similar to that of a composting toilet.

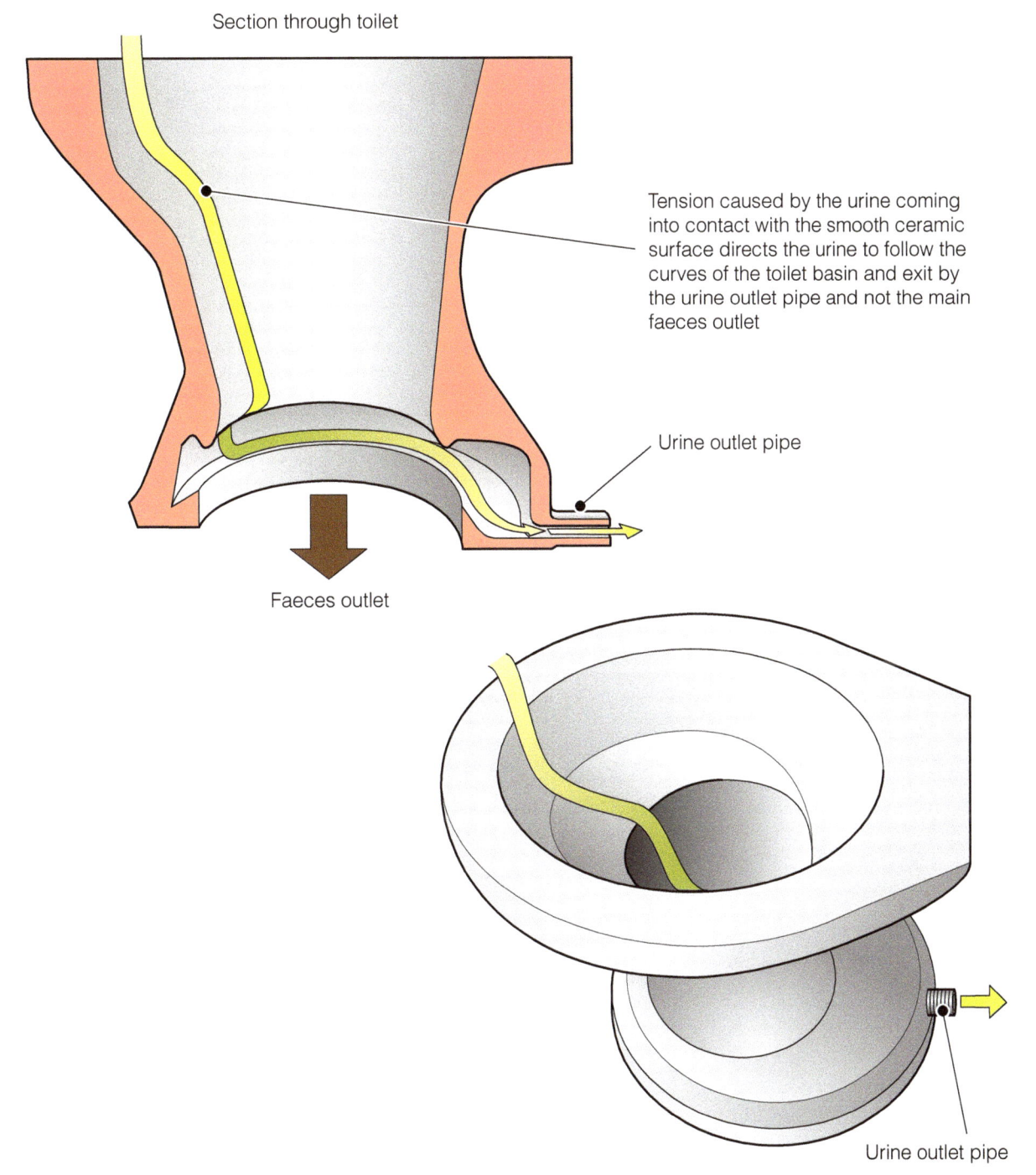

Section through toilet

Tension caused by the urine coming into contact with the smooth ceramic surface directs the urine to follow the curves of the toilet basin and exit by the urine outlet pipe and not the main faeces outlet

Urine outlet pipe

Faeces outlet

Urine outlet pipe

Links
http://www.otjitoilet.org/
http://www.youtube.com/watch?v=wvetp9F8G4U

PeePoo

Peepoo is a personal, single-use, self-sanitising, fully biodegradable toilet that prevents faeces from contaminating the immediate area as well as the surrounding ecosystem. After use, Peepoo turns into valuable fertiliser.

Bag ready for use

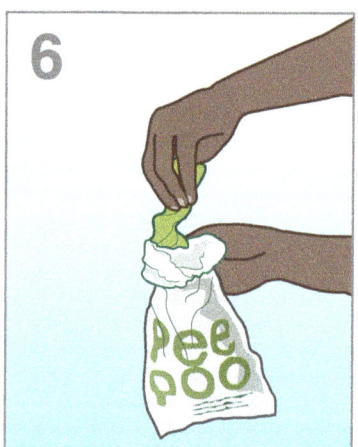

Bag is made from degradable bioplastic lined with urea, is safe to handle remaining odour-free for at least 24hrs

Cardboard and plastic toilets to hold the bag when in use, are also available

Links
http://www.youtube.com/watch?v=UJZhS252tdM
http://www.peepoople.com/

Contemporary Toilet Designs

Piet Urine Diversion Toilet*

Theo Brandwijk developed the Piet (pee plus seat) toilet that separates urine and faeces. This makes it cheaper to clean the sewage water, and the nitrogen and phosphates can be recycled to produce fertilizer. Piet works with a movement sensor, which opens a three-way valve. The urine is diverted down towards a special drain at the front leading to a tank that is emptied periodically. The faeces disappear through the larger regular drain. The three-way valve automatically closes when the user stands up.

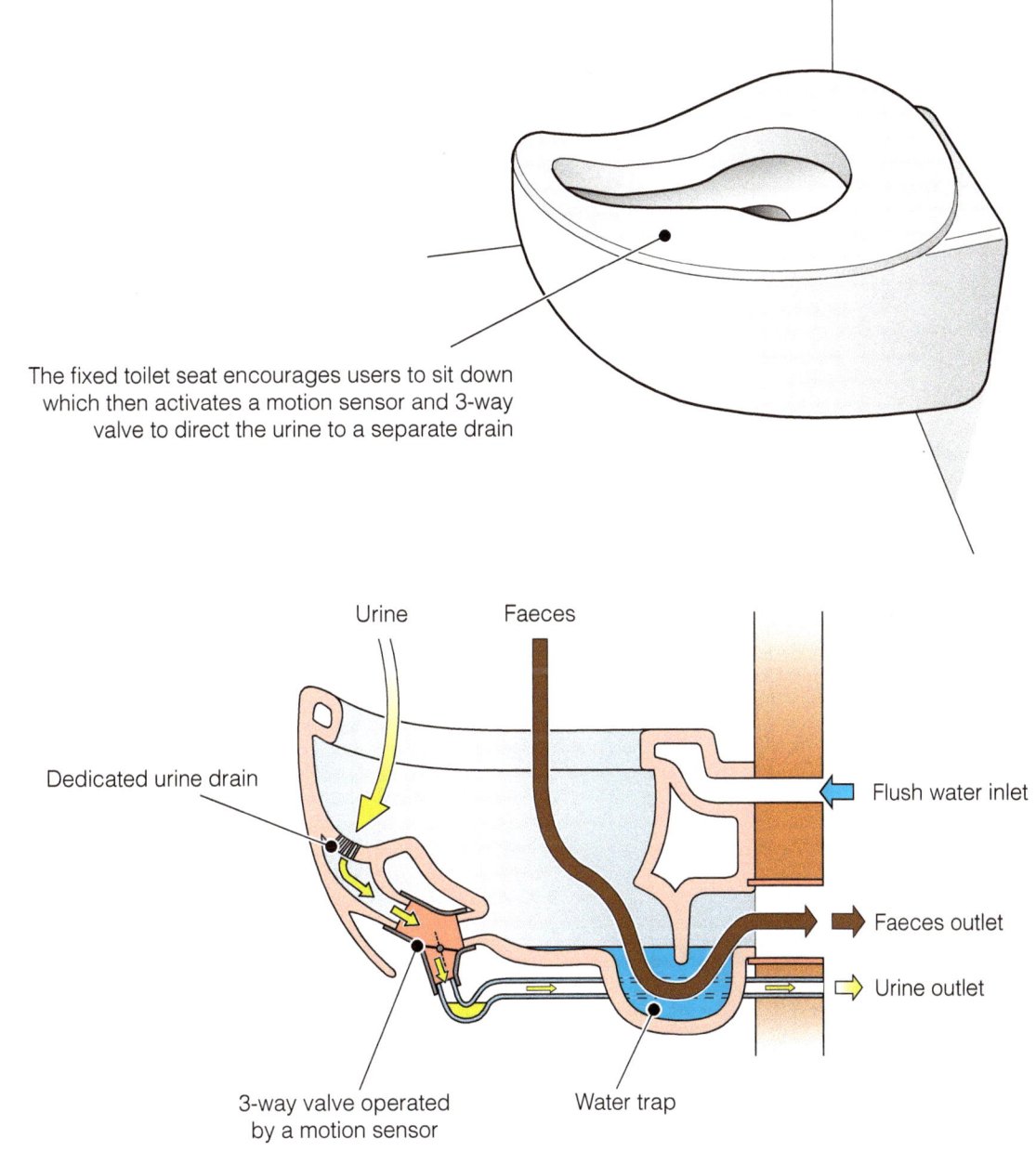

The fixed toilet seat encourages users to sit down which then activates a motion sensor and 3-way valve to direct the urine to a separate drain

Urine
Faeces
Dedicated urine drain
Flush water inlet
Faeces outlet
Urine outlet
3-way valve operated by a motion sensor
Water trap

Links
http://www.design.nl/item/the_problem_with_poo
typischtheo.nl/

Public Rest Room

Designed by Amos Bender, this is a sanitation and composting system for a public rest room (or rest area). The toilet room has four separate toilet basins that keep liquid waste apart from the solid waste and make it easy to add and mix a little soil with the solid waste to aid composting. In this particular composting system there are three separate heaps of waste material, each in a different stage of action spread over three years, with the compost being ready for use as a fertilzer in its third year. (See also pages 6 and 19.)

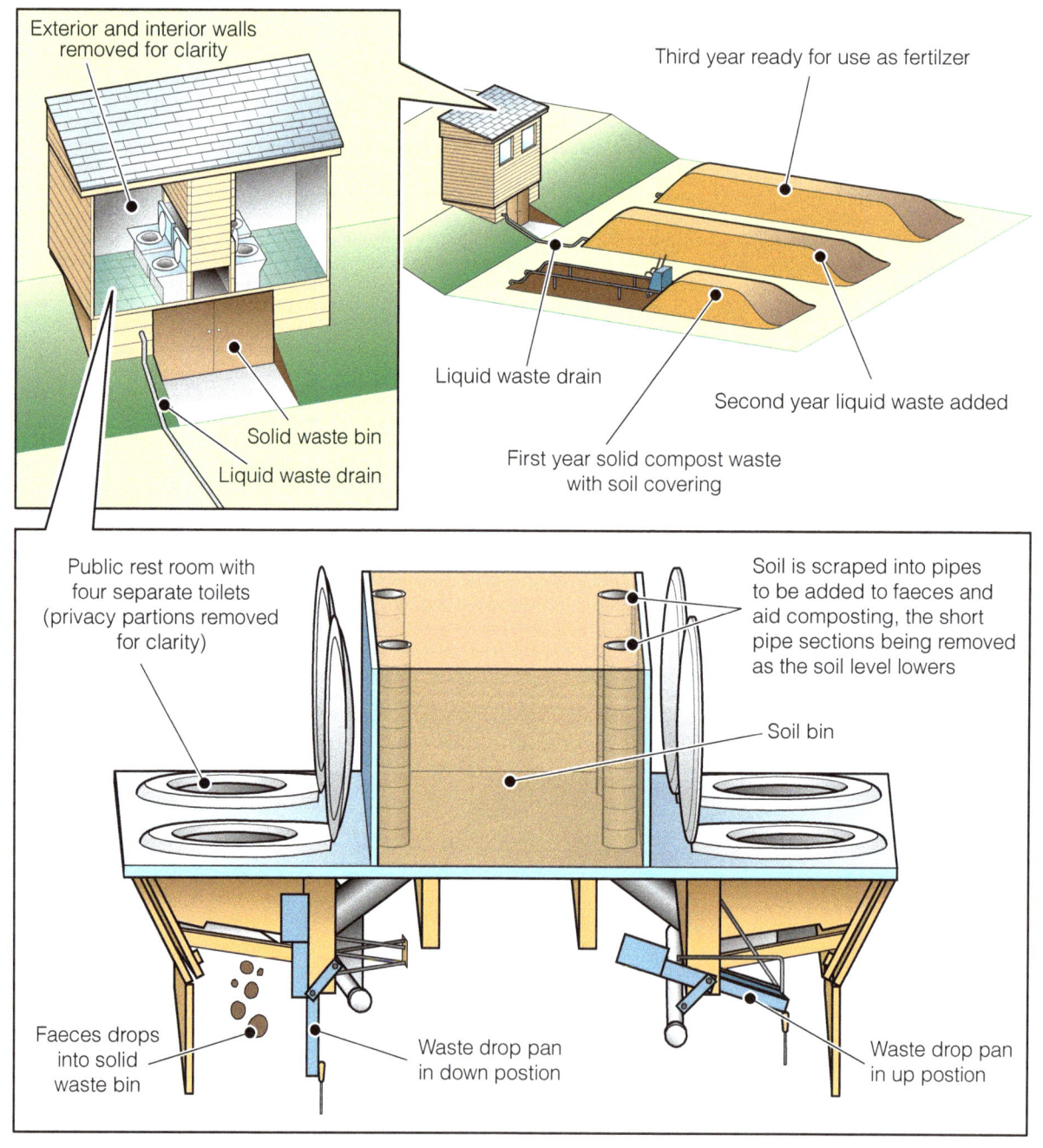

Links
http://www.exploringnaturespossibilities.com/

Contemporary Toilet Designs

Resource Toilet

The Resource is an ultra-low cost toilet with a removable container that makes it easy to collect and transport waste safely from the community. The toilet combines a 20 litre bucket, a liquid container, and a western-style toilet seat into a sealed, portable, urine diverting toilet.

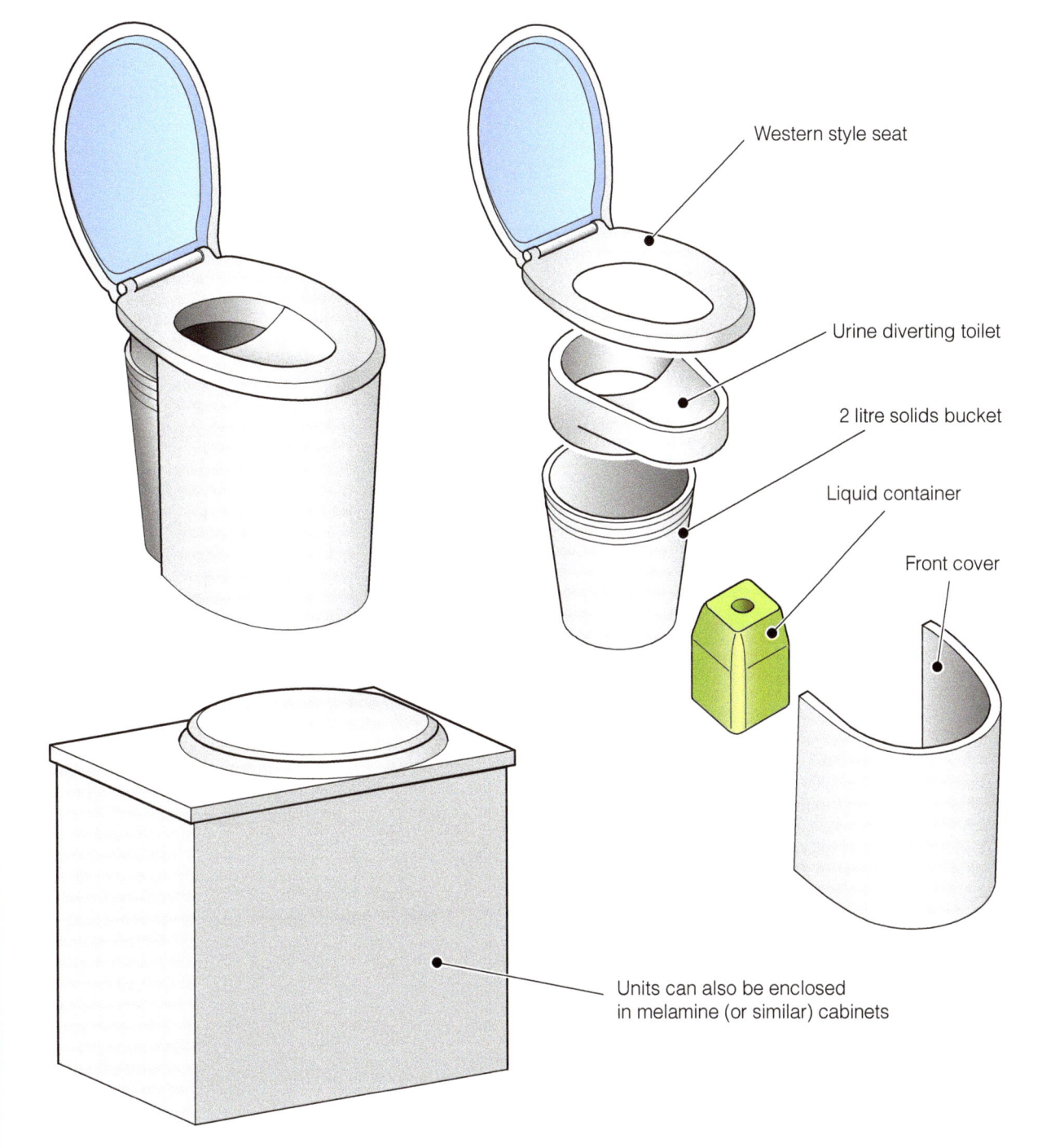

Western style seat

Urine diverting toilet

2 litre solids bucket

Liquid container

Front cover

Units can also be enclosed in melamine (or similar) cabinets

Links

http://www.resourcesanitation.com
http://resourcesanitation.com/2013/11/14/we-share-our-encyclopedia

Rolling Toilet: x-runner

Israel's Noa Lerner, a Berlin-based industrial engineer, designed this toilet for slum dwellers. It consists of a squatting platform, placed over a removable container that can be rolled to a neighbourhood collection facility.

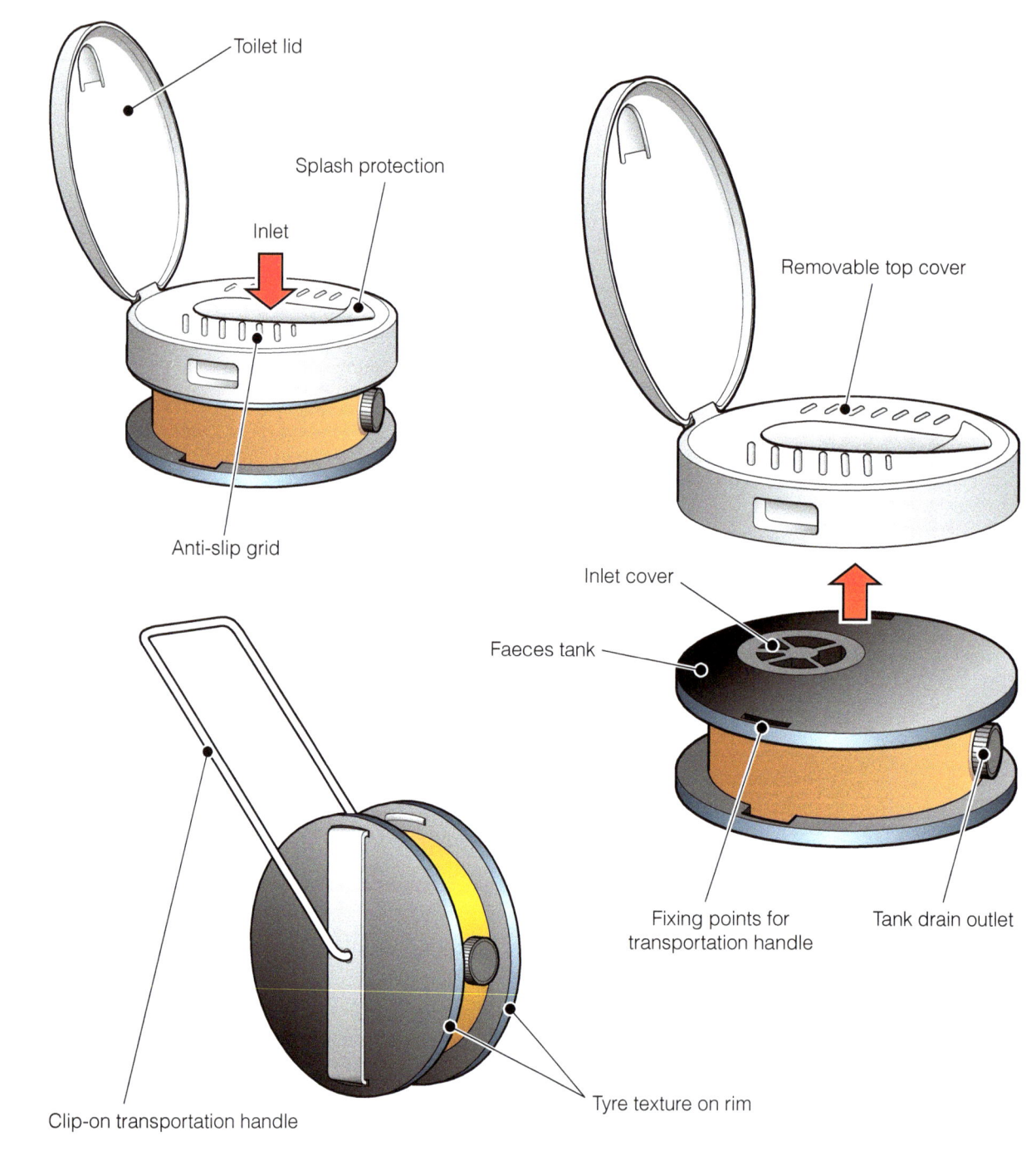

Links
http://www.youtube.com/watch?v=NSML-cG46Fo
http://www.xrunner-venture.com/
http://vimeo.com/51312933#
http://www.greenprophet.com/2011/04/israeli-designer-green-toilet-indias-slum-dwellers/

Contemporary Toilet Designs

SimSan

The SimSan BucketMate from Amos Bender is a simple toilet that makes human waste easy to manage as it can be used with a common bucket with 95% of urine being directed away to another location. The toilet incorporates a unique gate that only needs to be opened for defecating and remains closed at all other times, blocking odour. The bucket is also divided into four individual smaller sections. When one section is full the SimSan toilet is simply rotated to the next providing extra capacity before it requires emptying.

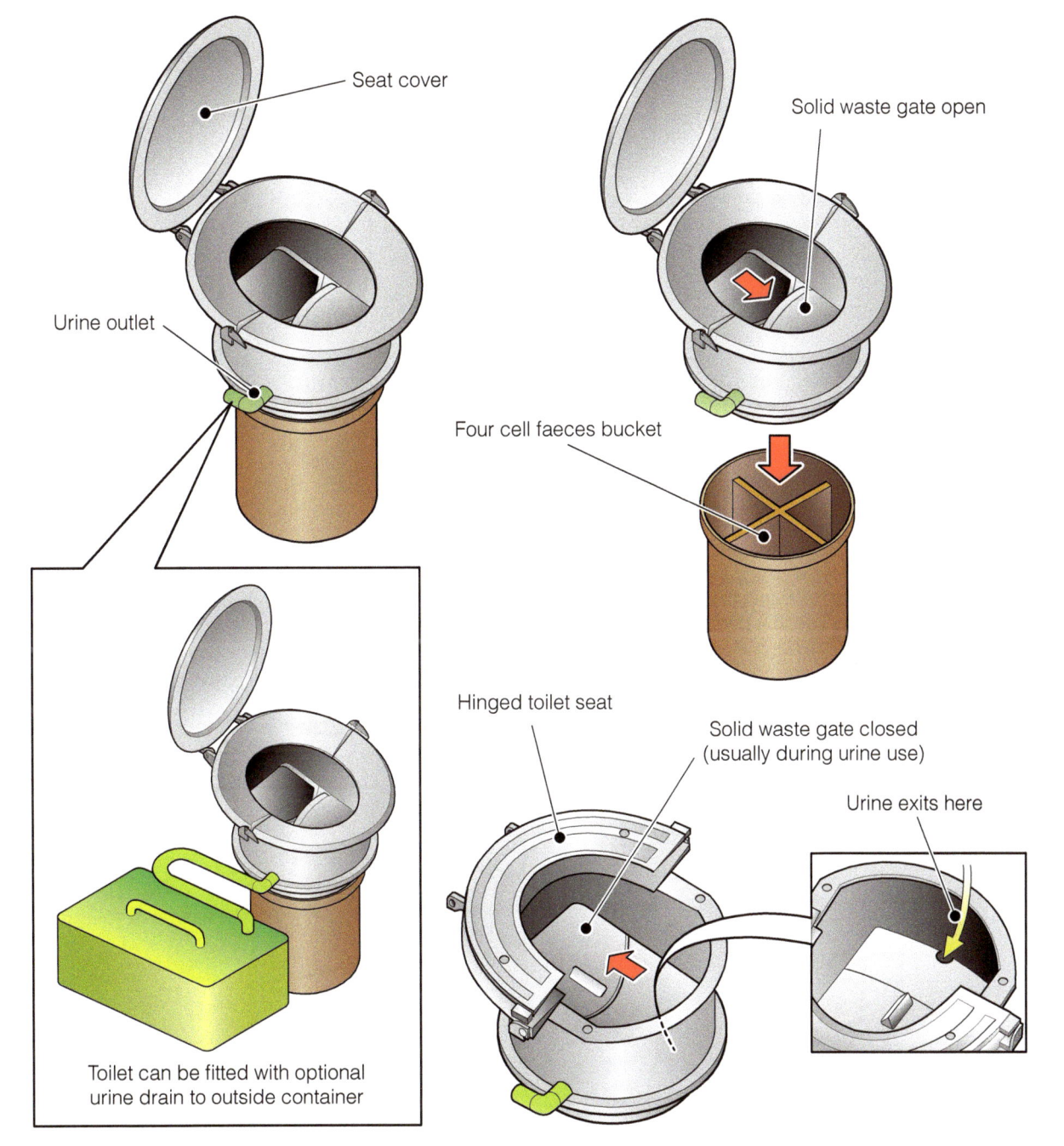

Links
http://www.liftupthepoor.net/SimSan.html

Waterless Toilet*

This student project was the second place winner of the Victorinox Time to Care, Sustainable Design Award in September 2011. It doesn't require water to function properly; doesn't generate sewage; is made for urban areas and generates compost suitable for crops and is user-friendly.

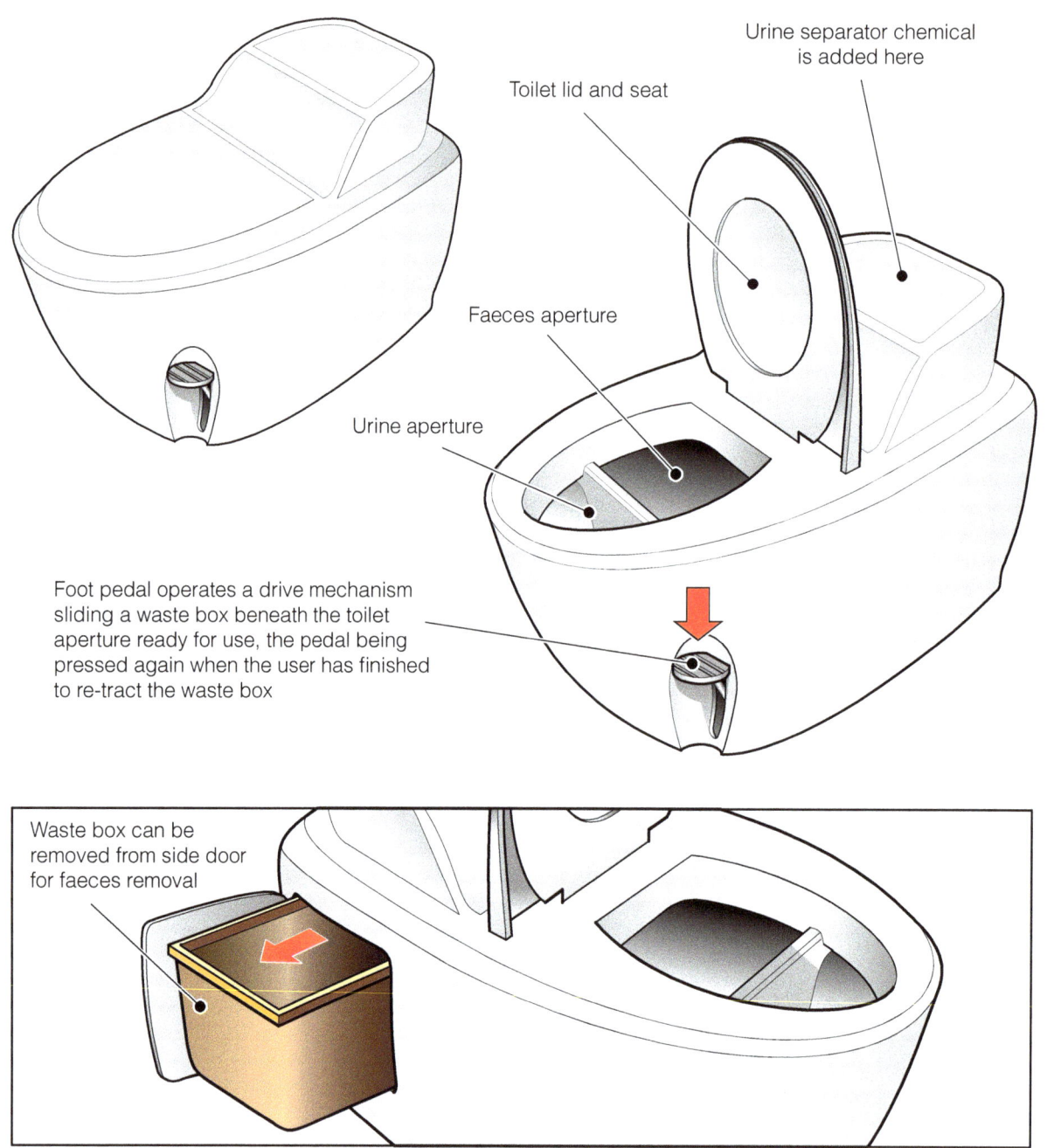

Urine separator chemical is added here

Toilet lid and seat

Faeces aperture

Urine aperture

Foot pedal operates a drive mechanism sliding a waste box beneath the toilet aperture ready for use, the pedal being pressed again when the user has finished to re-tract the waste box

Waste box can be removed from side door for faeces removal

Links

http://forum.susana.org/forum/categories/34-urine-diversion-systems-in-cludes-uddt-and-ud-flush-toilet/706-colombiamexico-aguayuda-is-looking-for-a-urinal-design

http://timetocare.victorinox.com/en/nc/vote-win/waterless-toilet.html

WC der Zukunft

Contemporary Toilet Designs

WC der Zukunft (Toilet of the Future) is a urine diversion composting toilet. Urine can be used either as a fertilzer or simply discharged into the sewage system. Faeces are collected with other recyclable materials and compost.

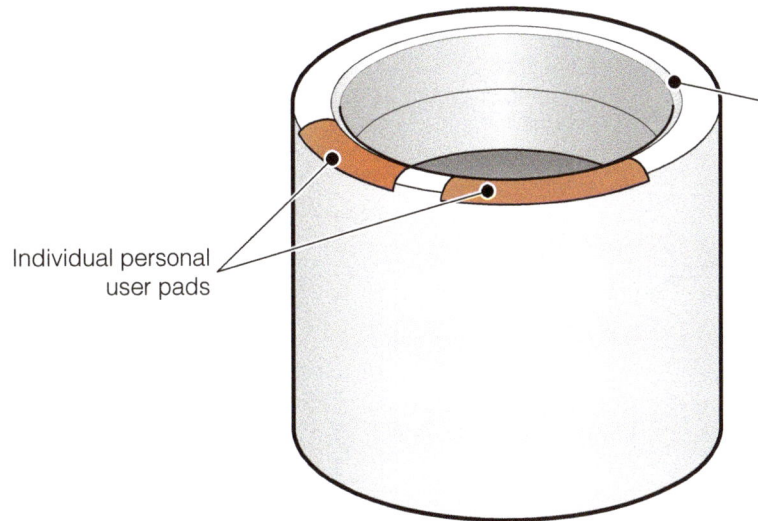

Individual personal user pads

Round toilet top reduces user contact surface to a minimum without loss of comfort and improves hygiene

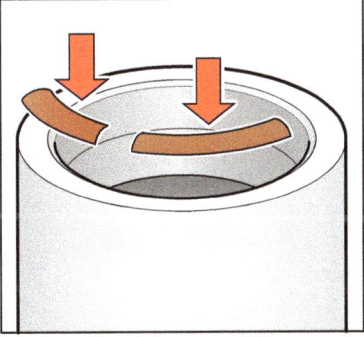

Each user places individual biomaterial pads on the toilet rim

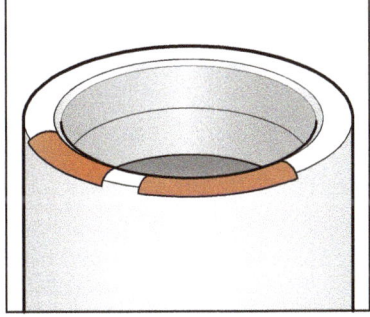

The pads increase user comfort and improve hygiene

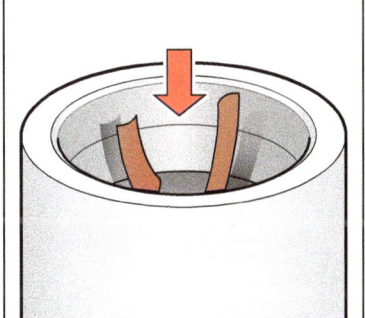

After use the pads are thrown into the toilet to aid composting

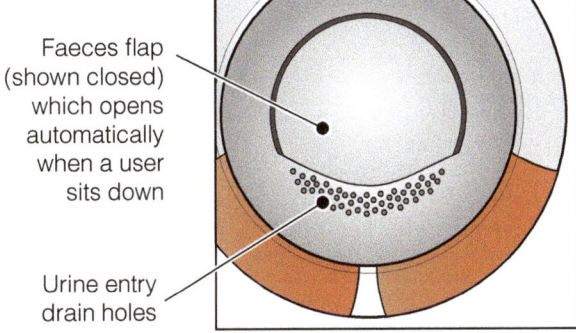

Faeces flap (shown closed) which opens automatically when a user sits down

Urine entry drain holes

Faeces flap in closed position

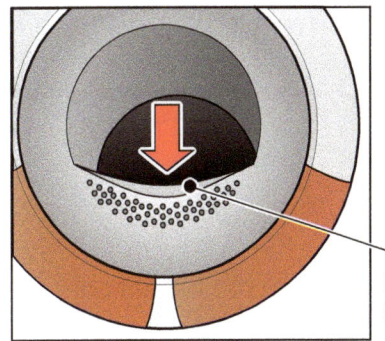

Faeces flap (shown open)

Toilet flap in open position

Links

http://www.ronja-scholz.de

WooWoo

Woo Woo is a London-based company, providing waterless toilets for sale throughout the UK. This free standing, public toilet system functions without water, electricity or chemicals and can be installed virtually anywhere. It offers an ideal solution for sites with no connection to water or sewerage services.

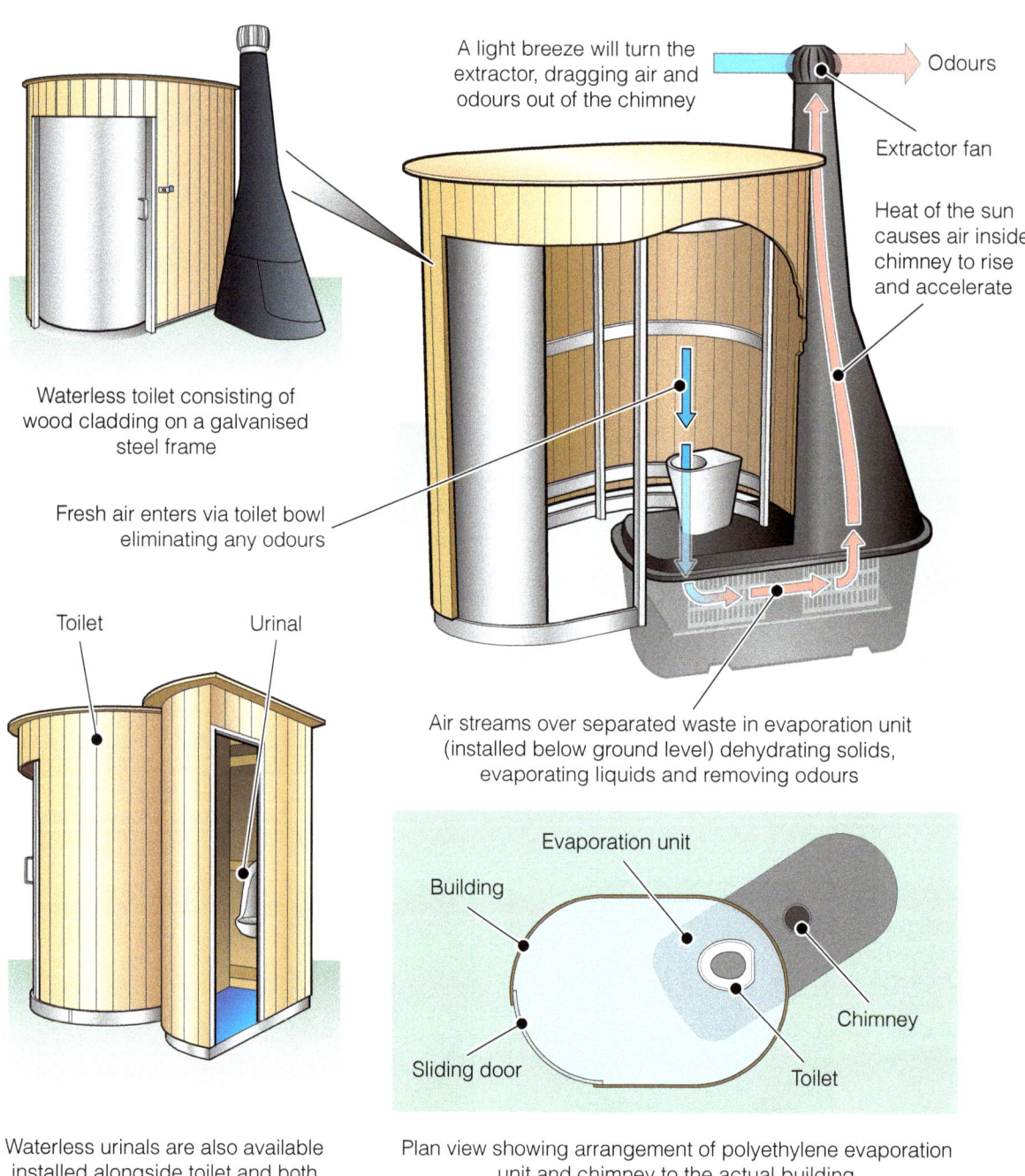

Waterless toilet consisting of wood cladding on a galvanised steel frame

Fresh air enters via toilet bowl eliminating any odours

A light breeze will turn the extractor, dragging air and odours out of the chimney

Odours

Extractor fan

Heat of the sun causes air inside chimney to rise and accelerate

Air streams over separated waste in evaporation unit (installed below ground level) dehydrating solids, evaporating liquids and removing odours

Waterless urinals are also available installed alongside toilet and both linked to the evaporation system

Plan view showing arrangement of polyethylene evaporation unit and chimney to the actual building

Links
http://www.waterlesstoilets.co.uk

Controlling Odour

Controlling odour from urine

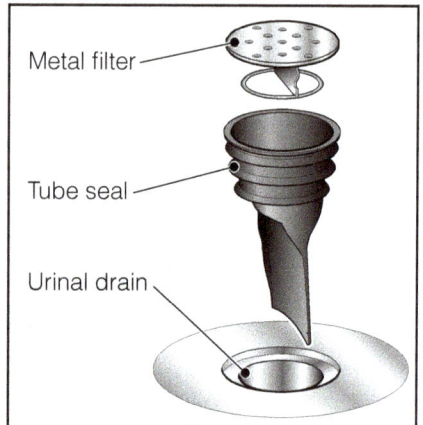

Rubber tube seal
This rubber tube is flat at the bottom when not in use (and hence blocks odour from the sewer or urine storage tank) but opens up when urine is flowing through. This one-way valve allows passage of grit up to 2mm.

It must be cleaned approximately once per month and replaced every year.

http://pro.keramag.com/produkte/urinale/centaurus.html

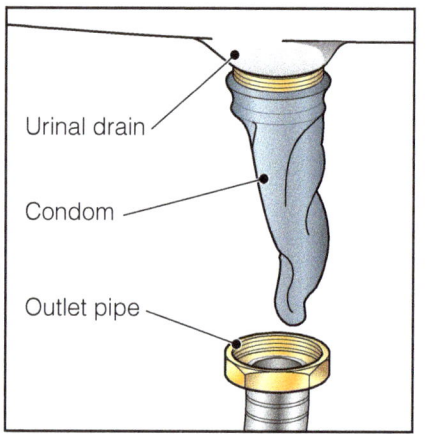

Condoms – The low-tech alternative
Condoms with the ends trimmed can be used as an effective seal for male urinals.

http://www.wecf.eu/download/2011/February/SSP-06_Jan2011_16-221.pdf

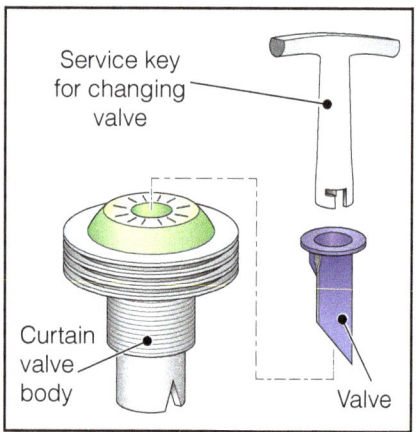

Valve seals
Valve seals are similar to the rubber tube seal, but are designed to reduce maintenance requirements. This seal has self-cleaning properties, although the production requires complex injection moulding machines, so it is not feasible to produce it locally.

http://www.autospec.co.za/productmedia/addicom/datasheets/addicomframe.htm
http://www.enswico.com/en/technology/key-system.html
http://www2.gtz.de/Dokumente/oe44/ecosan/en-waterless-urinals-a-proposal-to-save-water-and-recover-urine-nutrients-in-africa-2009.pdf
http://www.culu.eu/zubehoer

Contemporary Toilet Designs

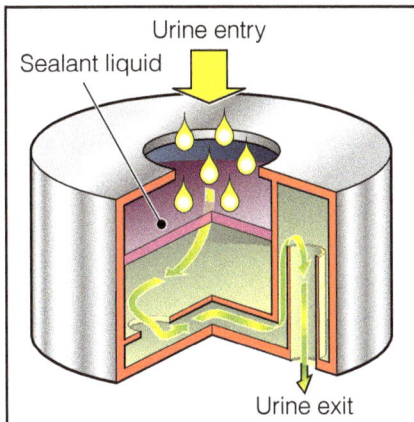

Sealant liquid

This system works with a sealant liquid (also called blocking fluid) which is made of vegetable oils or aliphatic alcohols. The sealant liquid, floats on top of the urine contained in the trap and thus constitutes an effective odour barrier. Urine immediately penetrates the sealant liquid and flows away to the drain.

http://www.freepatentsonline.com/6589440.html

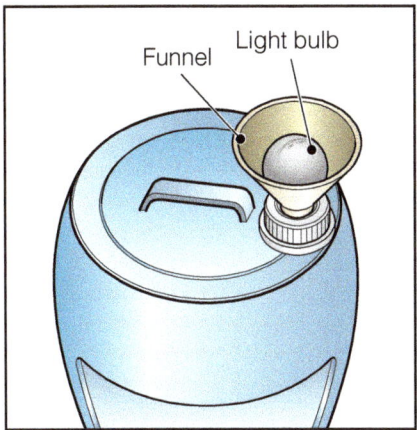

Light bulb / table tennis ball

Another low-tech solution: an old light bulb is placed in a funnel which in turn is inserted in the opening of a jerrican. This provides a portable and ready-made urinal. The contents should be clearly labelled on the outside of the can which must not be later used for water collection.

Controlling odour from faeces

Ventilation

Ventilated improved pit latrines (VIP latrines) can significantly reduce the smell of excreta and can even be more pleasant to use than some other water-based technologies. Flies that hatch in the pit are attracted to the light at the top of the ventilation pipe where they are trapped by a fly-screen and die.

The vent works better in windy areas but where there is little wind, its effectiveness can be improved by painting the pipe black; the heat difference between the pit (cool) and the vent (warm) creates an updraft that pulls the air and odours up and out of the pit.

Ventilating other types of latrine and keeping them clean will also help reduce the odour from faeces.

Soil, ash, sawdust, sand or lime

By covering the faeces with one of these dry materials, flies and odours are kept to a minimum.

http://www.eawag.ch/forschung/sandec/publikationen/compendium_e/index_EN

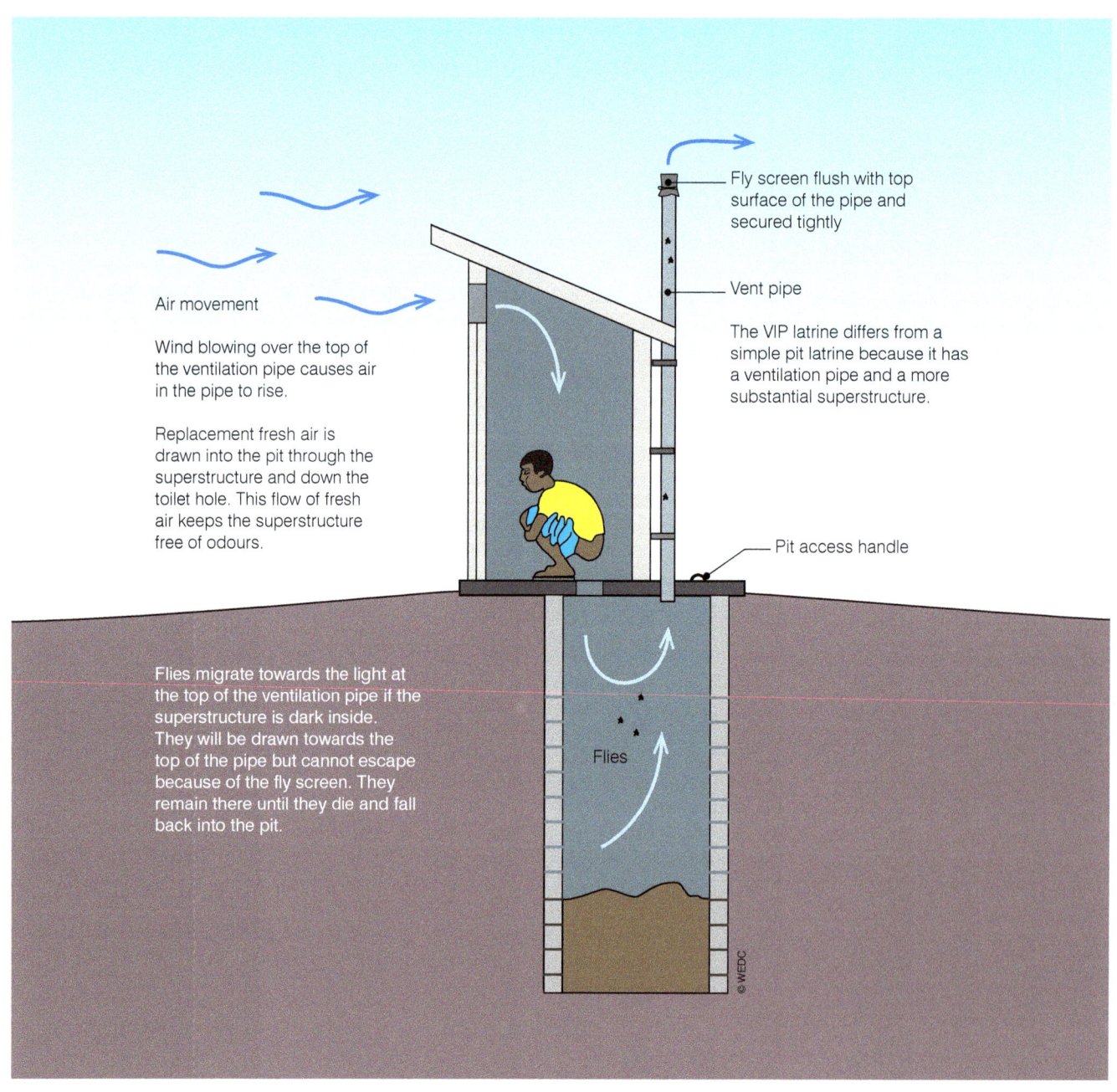

www.ingramcontent.com/pod-product-compliance
Ingram Content Group UK Ltd.
Pitfield, Milton Keynes, MK11 3LW, UK
UKHW050226150426
5217IPUK00018B/1270